CLASSICAL SEQUENCES IN BANACH SPACES

PURE AND APPLIED MATHEMATICS

A Program of Monographs, Textbooks, and Lecture Notes

MONOGRAPHS AND TEXTBOOKS IN
PURE AND APPLIED MATHEMATICS

1. *K. Yano,* Integral Formulas in Riemannian Geometry (1970)
2. *S. Kobayashi,* Hyperbolic Manifolds and Holomorphic Mappings (1970)
3. *V. S. Vladimirov,* Equations of Mathematical Physics (A. Jeffrey, editor; A. Littlewood, translator) (1970)
4. *B. N. Pshenichnyi,* Necessary Conditions for an Extremum (L. Neustadt, translation editor; K. Makowski, translator) (1971)
5. *L. Narici, E. Beckenstein, and G. Bachman,* Functional Analysis and Valuation Theory (1971)
6. *S. S. Passman,* Infinite Group Rings (1971)
7. *L. Dornhoff,* Group Representation Theory (in two parts). Part A: Ordinary Representation Theory. Part B: Modular Representation Theory (1971, 1972)
8. *W. Boothby and G. L. Weiss (eds.),* Symmetric Spaces: Short Courses Presented at Washington University (1972)
9. *Y. Matsushima,* Differentiable Manifolds (E. T. Kobayashi, translator) (1972)
10. *L. E. Ward, Jr.,* Topology: An Outline for a First Course (1972)
11. *A. Babakhanian,* Cohomological Methods in Group Theory (1972)
12. *R. Gilmer,* Multiplicative Ideal Theory (1972)
13. *J. Yeh,* Stochastic Processes and the Wiener Integral (1973)
14. *J. Barros-Neto,* Introduction to the Theory of Distributions (1973)
15. *R. Larsen,* Functional Analysis: An Introduction (1973)
16. *K. Yano and S. Ishihara,* Tangent and Cotangent Bundles: Differential Geometry (1973)
17. *C. Procesi,* Rings with Polynomial Identities (1973)
18. *R. Hermann,* Geometry, Physics, and Systems (1973)
19. *N. R. Wallach,* Harmonic Analysis on Homogeneous Spaces (1973)
20. *J. Dieudonné,* Introduction to the Theory of Formal Groups (1973)
21. *I. Vaisman,* Cohomology and Differential Forms (1973)
22. *B.-Y. Chen,* Geometry of Submanifolds (1973)
23. *M. Marcus,* Finite Dimensional Multilinear Algebra (in two parts) (1973, 1975)
24. *R. Larsen,* Banach Algebras: An Introduction (1973)
25. *R. O. Kujala and A. L. Vitter (eds.),* Value Distribution Theory: Part A; Part B: Deficit and Bezout Estimates by Wilhelm Stoll (1973)
26. *K. B. Stolarsky,* Algebraic Numbers and Diophantine Approximation (1974)
27. *A. R. Magid,* The Separable Galois Theory of Commutative Rings (1974)
28. *B. R. McDonald,* Finite Rings with Identity (1974)
29. *J. Satake,* Linear Algebra (S. Koh, T. A. Akiba, and S. Ihara, translators) (1975)
30. *J. S. Golan,* Localization of Noncommutative Rings (1975)
31. *G. Klambauer,* Mathematical Analysis (1975)
32. *M. K. Agoston,* Algebraic Topology: A First Course (1976)
33. *K. R. Goodearl,* Ring Theory: Nonsingular Rings and Modules (1976)
34. *L. E. Mansfield,* Linear Algebra with Geometric Applications: Selected Topics (1976)
35. *N. J. Pullman,* Matrix Theory and Its Applications (1976)
36. *B. R. McDonald,* Geometric Algebra Over Local Rings (1976)
37. *C. W. Groetsch,* Generalized Inverses of Linear Operators: Representation and Approximation (1977)
38. *J. E. Kuczkowski and J. L. Gersting,* Abstract Algebra: A First Look (1977)
39. *C. O. Christenson and W. L. Voxman,* Aspects of Topology (1977)
40. *M. Nagata,* Field Theory (1977)
41. *R. L. Long,* Algebraic Number Theory (1977)
42. *W. F. Pfeffer,* Integrals and Measures (1977)
43. *R. L. Wheeden and A. Zygmund,* Measure and Integral: An Introduction to Real Analysis (1977)
44. *J. H. Curtiss,* Introduction to Functions of a Complex Variable (1978)
45. *K. Hrbacek and T. Jech,* Introduction to Set Theory (1978)

46. *W. S. Massey,* Homology and Cohomology Theory (1978)
47. *M. Marcus,* Introduction to Modern Algebra (1978)
48. *E. C. Young,* Vector and Tensor Analysis (1978)
49. *S. B. Nadler, Jr.,* Hyperspaces of Sets (1978)
50. *S. K. Segal,* Topics in Group Kings (1978)
51. *A. C. M. van Rooij,* Non-Archimedean Functional Analysis (1978)
52. *L. Corwin and R. Szczarba,* Calculus in Vector Spaces (1979)
53. *C. Sadosky,* Interpolation of Operators and Singular Integrals: An Introduction to Harmonic Analysis (1979)
54. *J. Cronin,* Differential Equations: Introduction and Quantitative Theory (1980)
55. *C. W. Groetsch,* Elements of Applicable Functional Analysis (1980)
56. *I. Vaisman,* Foundations of Three-Dimensional Euclidean Geometry (1980)
57. *H. I. Freedan,* Deterministic Mathematical Models in Population Ecology (1980)
58. *S. B. Chae,* Lebesgue Integration (1980)
59. *C. S. Rees, S. M. Shah, and C. V. Stanojević,* Theory and Applications of Fourier Analysis (1981)
60. *L. Nachbin,* Introduction to Functional Analysis: Banach Spaces and Differential Calculus (R. M. Aron, translator) (1981)
61. *G. Orzech and M. Orzech,* Plane Algebraic Curves: An Introduction Via Valuations (1981)
62. *R. Johnsonbaugh and W. E. Pfaffenberger,* Foundations of Mathematical Analysis (1981)
63. *W. L. Voxman and R. H. Goetschel,* Advanced Calculus: An Introduction to Modern Analysis (1981)
64. *L. J. Corwin and R. H. Szcarba,* Multivariable Calculus (1982)
65. *V. I. Istrățescu,* Introduction to Linear Operator Theory (1981)
66. *R. D. Järvinen,* Finite and Infinite Dimensional Linear Spaces: A Comparative Study in Algebraic and Analytic Settings (1981)
67. *J. K. Beem and P. E. Ehrlich,* Global Lorentzian Geometry (1981)
68. *D. L. Armacost,* The Structure of Locally Compact Abelian Groups (1981)
69. *J. W. Brewer and M. K. Smith, eds.,* Emily Noether: A Tribute to Her Life and Work (1981)
70. *K. H. Kim,* Boolean Matrix Theory and Applications (1982)
71. *T. W. Wieting,* The Mathematical Theory of Chromatic Plane Ornaments (1982)
72. *D. B. Gauld,* Differential Topology: An Introduction (1982)
73. *R. L. Faber,* Foundations of Euclidean and Non-Euclidean Geometry (1983)
74. *M. Carmeli,* Statistical Theory and Random Matrices (1983)
75. *J. H. Carruth, J. A. Hildebrant, and R. J. Koch,* The Theory of Topological Semigroups (1983)
76. *R. L. Faber,* Differential Geometry and Relativity Theory: An Introduction (1983)
77. *S. Barnett,* Polynomials and Linear Control Systems (1983)
78. *G. Karpilovsky,* Commutative Group Algebras (1983)
79. *F. Van Oystaeyen and A. Verschoren,* Relative Invariants of Rings: The Commutative Theory (1983)
80. *I. Vaisman,* A First Course in Differential Geometry (1984)
81. *G. W. Swan,* Applications of Optimal Control Theory in Biomedicine (1984)
82. *T. Petrie and J. D. Randall,* Transformation Groups on Manifolds (1984)
83. *K. Goebel and S. Reich,* Uniform Convexity, Hyperbolic Geometry, and Nonexpansive Mappings (1984)
84. *T. Albu and C. Năstăsescu,* Relative Finiteness in Module Theory (1984)
85. *K. Hrbacek and T. Jech,* Introduction to Set Theory: Second Edition, Revised and Expanded (1984)
86. *F. Van Oystaeyen and A. Verschoren,* Relative Invariants of Rings: The Noncommutative Theory (1984)
87. *B. R. McDonald,* Linear Algebra Over Commutative Rings (1984)
88. *M. Namba,* Geometry of Projective Algebraic Curves (1984)
89. *G. F. Webb,* Theory of Nonlinear Age-Dependent Population Dynamics (1985)
90. *M. R. Bremner, R. V. Moody, and J. Patera,* Tables of Dominant Weight Multiplicities for Representations of Simple Lie Algebras (1985)

91. *A. E. Fekete,* Real Linear Algebra (1985)
92. *S. B. Chae,* Holomorphy and Calculus in Normed Spaces (1985)
93. *A. J. Jerri,* Introduction to Integral Equations with Applications (1985)
94. *G. Karpilovsky,* Projective Representations of Finite Groups (1985)
95. *L. Narici and E. Beckenstein,* Topological Vector Spaces (1985)
96. *J. Weeks,* The Shape of Space: How to Visualize Surfaces and Three-Dimensional Manifolds (1985)
97. *P. R. Gribik and K. O. Kortanek,* Extremal Methods of Operations Research (1985)
98. *J.-A. Chao and W. A. Woyczynski, eds.,* Probability Theory and Harmonic Analysis (1986)
99. *G. D. Crown, M. H. Fenrick, and R. J. Valenza,* Abstract Algebra (1986)
100. *J. H. Carruth, J. A. Hildebrant, and R. J. Koch,* The Theory of Topological Semigroups, Volume 2 (1986)
101. *R. S. Doran and V. A. Belfi,* Characterizations of C*-Algebras: The Gelfand-Naimark Theorems (1986)
102. *M. W. Jeter,* Mathematical Programming: An Introduction to Optimization (1986)
103. *M. Altman,* A Unified Theory of Nonlinear Operator and Evolution Equations with Applications: A New Approach to Nonlinear Partial Differential Equations (1986)
104. *A. Verschoren,* Relative Invariants of Sheaves (1987)
105. *R. A. Usmani,* Applied Linear Algebra (1987)
106. *P. Blass and J. Lang,* Zariski Surfaces and Differential Equations in Characteristic p > 0 (1987)
107. *J. A. Reneke, R. E. Fennell, and R. B. Minton,* Structured Hereditary Systems (1987)
108. *H. Busemann and B. B. Phadke,* Spaces with Distinguished Geodesics (1987)
109. *R. Harte,* Invertibility and Singularity for Bounded Linear Operators (1988)
110. *G. S. Ladde, V. Lakshmikantham, and B. G. Zhang,* Oscillation Theory of Differential Equations with Deviating Arguments (1987)
111. *L. Dudkin, I. Rabinovich, and I. Vakhutinsky,* Iterative Aggregation Theory: Mathematical Methods of Coordinating Detailed and Aggregate Problems in Large Control Systems (1987)
112. *T. Okubo, Differential Geometry* (1987)
113. *D. L. Stancl and M. L. Stancl,* Real Analysis with Point-Set Topology (1987)
114. *T. C. Gard,* Introduction to Stochastic Differential Equations (1988)
115. *S. S. Abhyankar,* Enumerative Combinatorics of Young Tableaux (1988)
116. *H. Strade and R. Farnsteiner,* Modular Lie Algebras and Their Representations (1988)
117. *J. A. Huckaba,* Commutative Rings with Zero Divisors (1988)
118. *W. D. Wallis,* Combinatorial Designs (1988)
119. *W. Więsław,* Topological Fields (1988)
120. *G. Karpilovsky,* Field Theory: Classical Foundations and Multiplicative Groups (1988)
121. *S. Caenepeel and F. Van Oystaeyen,* Brauer Groups and the Cohomology of Graded Rings (1989)
122. *W. Kozlowski,* Modular Function Spaces (1988)
123. *E. Lowen-Colebunders,* Function Classes of Cauchy Continuous Maps (1989)
124. *M. Pavel,* Fundamentals of Pattern Recognition (1989)
125. *V. Lakshmikantham, S. Leela, and A. A. Martynyuk,* Stability Analysis of Nonlinear Systems (1989)
126. *R. Sivaramakrishnan,* The Classical Theory of Arithmetic Functions (1989)
127. *N. A. Watson,* Parabolic Equations on an Infinite Strip (1989)
128. *K. J. Hastings,* Introduction to the Mathematics of Operations Research (1989)
129. *B. Fine,* Algebraic Theory of the Bianchi Groups (1989)
130. *D. N. Dikranjan, I. R. Prodanov, and L. N. Stoyanov,* Topological Groups: Characters, Dualities, and Minimal Group Topologies (1989)
131. *J. C. Morgan II,* Point Set Theory (1990)
132. *P. Biler and A. Witkowski,* Problems in Mathematical Analysis (1990)
133. *H. J. Sussmann,* Nonlinear Controllability and Optimal Control (1990)
134. *J.-P. Florens, M. Mouchart, and J. M. Rolin,* Elements of Bayesian Statistics (1990)
135. *N. Shell,* Topological Fields and Near Valuations (1990)
136. *B. F. Doolin and C. F. Martin,* Introduction to Differential Geometry for Engineers (1990)

137. *S. S. Holland, Jr.*, Applied Analysis by the Hilbert Space Method (1990)
138. *J. Okniński,* Semigroup Algebras (1990)
139. *K. Zhu,* Operator Theory in Function Spaces (1990)
140. *G. B. Price,* An Introduction to Multicomplex Spaces and Functions (1991)
141. *R. B. Darst,* Introduction to Linear Programming: Applications and Extensions (1991)
142. *P. L. Sachdev,* Nonlinear Ordinary Differential Equations and Their Applications (1991)
143. *T. Husain,* Orthogonal Schauder Bases (1991)
144. *J. Foran,* Fundamentals of Real Analysis (1991)
145. *W. C. Brown,* Matrices and Vector Spaces (1991)
146. *M. M. Rao and Z. D. Ren,* Theory of Orlicz Spaces (1991)
147. *J. S. Golan and T. Head,* Modules and the Structures of Rings: A Primer (1991)
148. *C. Small,* Arithmetic of Finite Fields (1991)
149. *K. Yang,* Complex Algebraic Geometry: An Introduction to Curves and Surfaces (1991)
150. *D. G. Hoffman, D. A. Leonard, C. C. Lindner, K. T. Phelps, C. A. Rodger, and J. R. Wall,* Coding Theory: The Essentials (1991)
151. *M. O. González,* Classical Complex Analysis (1992)
152. *M. O. González,* Complex Analysis: Selected Topics (1992)
153. *L. W. Baggett,* Functional Analysis: A Primer (1992)
154. *M. Sniedovich,* Dynamic Programming (1992)
155. *R. P. Agarwal,* Difference Equations and Inequalities: Theory, Methods, and Applications (1992)
156. *C. Brezinski,* Biorthogonality and Its Applications to Numerical Analysis (1992)
157. *C. Swartz,* An Introduction to Functional Analysis (1992)
158. *S. B. Nadler, Jr.,* Continuum Theory: An Introduction (1992)
159. *M. A. Al-Gwaiz,* Theory of Distributions (1992)
160. *E. Perry,* Geometry: Axiomatic Developments with Problem Solving (1992)
161. *E. Castillo and M. R. Ruiz-Cobo,* Functional Equations and Modelling in Science and Engineering (1992)
162. *A. J. Jerri,* Integral and Discrete Transforms with Applications and Error Analysis (1992)
163. *A. Charlier, A. Bérard, M.-F. Charlier, and D. Fristot,* Tensors and the Clifford Algebra: Application to the Physics of Bosons and Fermions (1992)
164. *P. Biler and T. Nadzieja,* Problems and Examples in Differential Equations (1992)
165. *E. Hansen,* Global Optimization Using Interval Analysis (1992)
166. *S. Guerre-Delabrière,* Classical Sequences in Banach Spaces (1992)

Additional Volumes in Preparation

CLASSICAL SEQUENCES IN BANACH SPACES

Sylvie Guerre-Delabrière

University of Paris VI
Paris, France

Marcel Dekker, Inc.   **New York • Basel • Hong Kong**

Library of Congress Cataloging-in-Publication Data

Guerre-Delabrière, Sylvie,
    Classical sequences in Banach spaces / Sylvie Guerre-Delabrière.
      p.   cm. — (Monographs and textbooks in pure and applied
mathematics)
    Includes bibliographical references (p.   ) and index.
    ISBN 0-8247-8723-4
    1. Banach spaces. 2. Sequence spaces. I. Title. II. Series.
QA322.2.G78   1992
515′.732—dc20   92-19239
                                             CIP

This book is printed on acid-free paper.

MARCEL DEKKER, INC.
270 Madison Avenue, New York, New York 10016

Current printing (last digit):
10 9 8 7 6 5 4 3 2 1

PRINTED IN THE UNITED STATES OF AMERICA

Foreword

This book deals with the following important aspect of modern analysis: Which Banach spaces contain almost isometric copies of one of the classical sequence spaces c_0 or ℓ_p for some $1 \le p < \infty$? (All Banach spaces here are taken to be infinite dimensional.) A remarkable discovery of B. S. Tsirelson, in 1974, yields that there are Banach spaces that contain no isomorphic copy of any of these spaces, while a recent discovery of E. Odell and T. Schlumprecht asserts that Hilbert space (i.e., ℓ_2) itself can be renormed to fail to have this property; that is, the renormed space fails to contain an almost isometric copy of ℓ_2. A deep result of D. Aldous, in 1980, proved that every Banach space that isometrically embeds in L_p (for some $1 \le p < \infty$) *has* this property, and subsequently J. L. Krivine and B. Maurey, in 1981, crystallized the important class of stable Banach spaces, generalizing Aldous' proof, to show these also contain almost isometric copies of ℓ_p for some $1 \le p < \infty$. It is a nontrivial result that L_p is stable, and moreover, $L_p(X)$ is stable, for $1 \le p < \infty$ and X a stable space. Thus every subspace of $L_p(L_q(L_r \cdots))$ contains an almost isometric copy of some ℓ_s. It remains an open question, however, if every subspace of a quotient of L_p con-

iii

tains an almost isometric (or just isomorphic) copy of some ℓ_r, for $p \neq 2$, $1 < p < \infty$.

The proofs of these results involve methods and ideas that are quite different from those usually employed in modern analysis. These methods are exposed here in complete detail. Moreover, several other deep discoveries are also presented, such as the existence of subspaces of ℓ_p without a basis (done in the case $p > 2$), Zippin's theorem characterizing ℓ_p spaces, Krivine's theorem, and the Guerre–Levy theorem characterizing the smallest r such that ℓ_r embeds in a general subspace of L_p.

These remarkable results on stable spaces have heretofore appeared only in research monographs. Sylvie Guerre-Delabrière has made a welcome contribution to the current scientific literature by rendering these results accessible in her timely book.

Haskell P. Rosenthal
University of Texas
Austin, Texas

Preface

Complete normed spaces, introduced by S. Banach in 1930, were expected to be very "regular": several structure theorems, true for all Banach spaces, were proved from 1930 to 1970, for example, by S. Banach himself [Ban], S. Banach and H. Steinhaus [Be3], A. Dvoretsky [Dv], A. Grothendieck. But in 1973, P. Enflo [E1] gave counterexamples to Banach's question: Does every Banach space have a basis? That showed that things were not as nice as expected. From then on, each time one thought to prove a structure theorem that would have showed more unity and more simplicity among Banach spaces, it happened to be false.

In this book we address the following questions:

Does every Banach space contain a subspace isomorphic to c_0 or ℓ_p for some $p \in [1, +\infty[$?

If a Banach space contains c_0 or ℓ_p for some p, does it contain it almost isometrically?

The second question is known as the *distortion problem*.

In 1974, B. S. Tsirelson constructed a counterexample to the first question, namely, a reflexive and separable Banach space not containing any subspace isomorphic to c_0 or ℓ_p for any $p \in [1, +\infty[$, [T], and so the answer to this question is negative. However, shortly after Tsirelson's example, some rich natural families of Banach spaces were discovered for which an affirmative answer was found, for example, stable Banach spaces, defined by J. L. Krivine and B. Maurey [KM].

Very recently, E. Odell and T. Schlumprecht proved that there is a renorming of ℓ_2 that does not contain a subspace almost isometric to ℓ_2. The answer to the distortion problem is thus also negative in general. On the contrary, by R. C. James' results [J4], the distortion problem has a positive answer when the subspace is c_0 or ℓ_1. Moreover, in stable Banach spaces, it is shown in [KM] that the general problem has a positive answer. So, as shown by the complexity of known results, these problems remain deep and fundamental open questions.

In this book, we give as wide as possible a survey of results and techniques related to these problems in Banach spaces. Beside the principal problems, we can mention also some questions that arise naturally in this field and describe what can be expected:

Does any Banach space X contain a subspace isomorphic to c_0, ℓ_1, or a reflexive space?

This question has a positive answer in the case that X has an unconditional basis. However, in 1992 T. Gowers constructed a counterexample in the general case.

Is it possible to characterize those p such that ℓ_p embeds in a given Banach space X?

By a result of the author and M. Levy [GLe], this can be done in the case where X is an L_p-space.

Is ℓ_p ($1 \leq p < +\infty$) or c_0 always finitely representable in Banach spaces?

A famous result of J. L. Krivine shows that the answer to this question is positive [K2].

Does every Banach space contain an unconditional basic sequence?

Quite recently, T. Gowers and B. Maurey independently found a counterexample to this last problem.

Let us give a detailed description of the contents of this book. All the above questions deal with "regular" structures in Banach spaces.

So it is natural to start this study with the notion of Schauder basis in Banach spaces. In Chapter I we give the main definitions about bases, and then we investigate the relations between bases of a space and its dual space. This finds an application in the case of unconditional bases; in that context one gets first results on Banach spaces containing ℓ_1 or c_0. In the same section, we present James' results on the distortion problem for ℓ_1 and c_0 and Rosenthal's characterization of spaces containing ℓ_1. Next, we prove that subspaces of ℓ_p, $(1 \leq p < +\infty)$ and c_0 always contain a copy of the whole space. We finish this chapter with two famous constructions: James' sequence, which characterizes non-reflexive spaces, and the Tsirelson space, which contains no ℓ_p or c_0.

In Chapter II we present the concepts of ultrapowers and spreading models in Banach spaces: it is easier to find ℓ_p or c_0 in an ultrapower of a space X or as a spreading model of X than in the space itself. Moreover, in this case, ℓ_p or c_0 is finitely representable in X. After giving definitions and main properties of these spaces, we study spaces that have ℓ_p or c_0 isometrically as a spreading model. Then we prove J. L. Krivine's theorem: ℓ_p $(1 \leq p < +\infty)$ or c_0 is block-finitely representable in any basic sequence of any Banach space.

In Chapter III we define stable Banach spaces. After giving basic properties of these spaces and showing that L_p-spaces are stable, we prove the main result for which they were introduced by J. L. Krivine and B. Maurey: they always contain ℓ_p for some $p \in [1, +\infty[$. Moreover, this embedding is "almost isometric." Next, we show that the stable Banach spaces are weakly sequentially complete.

The first three sections of Chapter IV are composed of useful functional analytic properties of L_p-spaces. First we reduce the study of all these spaces to the case of ℓ_p and $L_p[0,1]$; then we prove some theorems related to the notion of p-equi-integrability, for instance, the so-called "subsequence splitting lemma," due to H. P. Rosenthal. We finish these sections with a description of ultrapowers of L_p-spaces. In the last two sections of this chapter, we characterize ℓ_q subspaces of L_p, and, for a given subspace X of L_p, we prove that $\ell_{p(X)}$ embeds in X, where $p(X)$ is the least upper bound of the set of p such that X is of Rademacher type p.

This book is intended for graduate students, and was used for a course at the University of Paris. It requires only a basic knowledge of classical functional analysis and the theory of Banach spaces; it is mostly self-contained. To help the reader, we recommend some references on functional analysis: [Ban], [Be3], [Bo1], [Bo2], [Br], [D], [DS], [Y]. Certain fundamental results of this book and further developments can be found in: [Be3], [BeL], [Di1], [Di2], [LT], [MiS], [Si].

I would like to thank Bernard Beauzamy, who read the first version of the manuscript and gave me much advice in improving the text.

Sylvie Guerre-Delabrière

Contents

Foreword   iii

Preface   v

Notation and Conventions   xi

I.   Classical Theorems   1

    1.   Schauder bases   1
    2.   Bases and duality   17
    3.   Unconditional bases   22
    4.   Banach spaces containing ℓ_1 or c_0 35
    5.   Subspaces of ℓ_p, $1 \leq p < +\infty$, or c_0 51
    6.   Reflexive spaces and James sequence   58
    7.   Tsirelson's space   63
    Notes and remarks   66
    Exercises   66

II.   Ultrapowers and Spreading Models   71

    1.   Finite representability and ultrapowers   72
    2.   Spreading models   78
    3.   Properties of I.S. sequences   83
    4.   Spreading models isometric to ℓ_p
        ($1 \leq p < \infty$) or c_0 90
    5.   Krivine's theorem   92
    Notes and remarks   103
    Exercises   104

III.   Stable Banach Spaces   108

    1.   Definitions, examples, and main tools   108
    2.   Characterizations of stable Banach spaces   113
    3.   Spreading models on stable spaces   120
    4.   The existence of ℓ_p subspaces   122
    5.   Other properties of stable spaces   127
    Notes and remarks   129
    Exercises   130

IV.   Subspaces of L_p-Spaces, $1 \leq p < +\infty$ 133

    1.   Reduction to the case of ℓ_p and $L_p[0,1]$ 133
    2.   Basic lemmas in $L_p[0,1]$, $1 \leq p < +\infty$ 139
    3.   Ultrapowers of L_p-spaces   153
    4. ℓ_q subspaces of $L_p[0,1]$ 161
    5. ℓ_q's in subspaces of $L_p[0,1]$ 171
    Notes and remarks   183
    Exercises   183

References   189

Index   203

Notation and Conventions

In this book, all spaces are real *Banach* spaces and all subspaces are *closed* subspaces. $X, Y, \ldots$ denote Banach spaces, $X^*, Y^*, \ldots$ their dual spaces.

We always consider X as a subspace of its second dual X^{**}.

$\| \cdot \|_X$ denotes the norm in X, or more simply $\| \cdot \|$ if there is no ambiguity.

$B_X, \overline{B}_X, S_X$ denote the open and closed unit balls of X and the unit sphere of X, respectively.

Weak topologies on X or X^* are denoted by $\sigma(X, X^*)$ and $\sigma(X^*, X^{**})$, and the weak* topology of X^* by $\sigma(X^*, X)$.

More generally, if $\mathcal{T}$ is a topology on X, a sequence $(x_n)_{n \in \mathbb{N}}$ will be $\mathcal{T}$-convergent (respectively, $\mathcal{T}$-Cauchy) if it is convergent for the topology $\mathcal{T}$ (respectively, Cauchy for the topology $\mathcal{T}$). If $\mathcal{T}$ is the topology given by the norm of X, we will just say convergent or Cauchy.

If Y is a (closed) subspace of a Banach space X, we write $Y \subset X$.

Linear continuous maps between two Banach spaces X and Y are called operators and denoted $T : X \to Y$. The transpose of T is denoted by $T^* : Y^* \to X^*$. If T is an isomorphism between X and Y, the isomorphism constant C is defined by:

$$C = \| T \| \| T^{-1} \|$$

and we write in this situation $X \overset{c}{\sim} Y$, or more simply $X \sim Y$ if we do not want to specify the isomorphism constant. We will say that X and Y are C-isomorphic or simply isomorphic. When Y is isomorphic to a subspace of X, we write $Y \hookrightarrow X$ or $Y \overset{c}{\hookrightarrow} X$ if C is the isomorphism constant.

The Banach–Mazur distance between two isomorphic Banach spaces X and Y is defined as:

$$d(X, Y) = \text{Inf}\{\| T \| \| T^{-1} \| / T : X \to Y \text{ is an isomorphism}\}$$

(d is not a distance but $\log d$ is.)

If E is a set in X, $\text{Span}[E]$ and $\overline{\text{Span}}[E]$ denote the linear span and the closed linear span of E in X, $\text{Conv}[E]$ and $\overline{\text{Conv}}[E]$ the convex hull and the closed convex hull of E. $\text{Diam}[E]$ denotes the diameter of E, that is, $\text{Sup}\{\| x - y \|, x, y \in E\}$. If E and E' are two subsets of X, $d(E, E')$ denotes the distance of E and E' associated with the norm of X, that is, $\text{Inf}\{\| x - y \|, x \in E, y \in E'\}$.

Let $(x_n)_{n \in \mathbb{N}}$ be a sequence in a Banach space X and let x be an element of X. We say that the series $\sum_{i=0}^{\infty} x_i$ converges to x whenever $\lim_{N \to +\infty} \| x - \sum_{i=0}^{N} x_i \| = 0$, and we write $x = \sum_{i=0}^{\infty} x_i$.

If E is a given set, $E^{(\mathbb{N})}$ denotes the set of finite sequences of elements of E, $E^{\mathbb{N}}$ the set of all sequences of elements of E.

If K is a compact set, $C(K)$ is the space of real-valued continuous functions on K with the Sup norm, that is:

$$\| f \|_\infty = \text{Sup}\{|f(t)|, t \in K\}.$$

For $1 \le p < +\infty$, ℓ_p denotes the closure of $\mathbb{R}^{(\mathbb{N})}$ under the norm:

$$\| (a_i)_{i \in \mathbb{N}} \|_p = \left(\sum_{i=0}^{\infty} |a_i|^p \right)^{1/p},$$

c_0 is the closure of $\mathbb{R}^{(\mathbb{N})}$ under the norm:

$$\| (a_i)_{i \in \mathbb{N}} \|_\infty = \text{Sup}_{i \in \mathbb{N}} |a_i| ;$$

ℓ_∞ is the space of bounded sequences of real numbers equipped with the norm $\| \cdot \|_\infty$.

For $1 \leq p \leq +\infty$, let ℓ_p^n be the space $\mathbb{R}^n$ equipped with the norm $\| \cdot \|$.

If (Ω, Σ, μ) is a measure space and $1 \leq p \leq +\infty$, $L_p(\Omega, \Sigma, \mu)$ is the space of classes of Σ-measurable functions on Ω such that:

$$\| f \| = \left(\int_\Omega |f(\omega)|^p \, d\mu(\omega) \right)^{1/p} < +\infty$$

(respectively, $\| f \|_\infty = \text{Ess-Sup}_{\omega \in \Omega} |f(\omega)| < +\infty$ if $p = +\infty$) equipped with the norm $\| \cdot \|_p$ (respectively, $\| \cdot \|_\infty$).

In the case where $\Omega = [0, 1]$, Σ is the Borel σ-field and μ is Lebesgue measure λ on $[0, 1]$, we just write $L_p[0, 1]$.

If X is a Banach space, $L_p(X) = L_p(\Omega, \Sigma, \mu; X)$ denotes the space of classes of X-valued Bochner Σ-measurable functions on Ω such that:

$$\| f \| = \left(\int_\Omega \| f(\omega) \|_X^p \, d\mu(\omega) \right)^{1/p} < +\infty$$

(respectively, $\| f \| = \text{Ess-Sup}_{\omega \in \Omega} \| f(\omega) \|_X < +\infty$ if $p = +\infty$).

As usual we do not distinguish between Σ and Σ/N where N is the σ-ideal of null sets in Σ (that is to say, $A \in N$ if $\mu(A) = 0$).

$X \oplus Y$ denotes the cartesian product of two Banach spaces X and Y, when the norm is not specified; $X \oplus_p Y$, $X \oplus_\infty Y$ are the cartesian products of X and Y equipped with the $\| \cdot \|_p$-norm or the $\| \cdot \|_\infty$-norm (i.e., $\| (x, y) \|_p = (\| x \|_X^p + \| y \|_Y^p)^{1/p}$, $\| (x, y) \|_\infty = \text{Sup}(\| x \|_X, \| y \|_Y)$). Moreover $(\oplus_p X_n)_{n \in \mathbb{N}}$ or $(X_0 \oplus X_1 \oplus \cdots \oplus X_n \oplus \cdots)_p$ is the closure of all finite sequences $(x_n)_{n \in \mathbb{N}}$ with $x_n \in X_n$ for all $n \in \mathbb{N}$ under the norm:

$$\| (x_n)_{n \in \mathbb{N}} \| = \left(\sum_{i=0}^\infty \| x_i \|_{X_i}^p \right)^{1/p} ;$$

$(\oplus_X X_n)_{n \in \mathbb{N}} = (X_0 \oplus X_1 \oplus \cdots)_X$ is defined similarly when X is a space with a basis $(e_n)_{n \in \mathbb{N}}$, the norm being defined by:

$$\| (x_n)_{n \in \mathbb{N}} \| = \left\| \sum_{i=0}^\infty \| x_i \|_{X_i} e_i \right\|_X.$$

Let us denote as usual $|I|$ the cardinality of a set I.

As stated in the Preface, in most of this book, we are going to work with real Banach spaces. Occasionally we must use complex spaces but we specify these cases very clearly. However, most of the results of this book extend with little change to the complex case.

We chose to present almost no results on Banach lattices except when we could not avoid them. There are results on lattices that are closely related to the subject of this book but the techniques are different and it would need another volume to describe them. The interested reader can see [LT], Vol II for a first approach to lattices.

CLASSICAL SEQUENCES
IN BANACH SPACES

I

Classical Theorems

1. SCHAUDER BASES

The notion of a Schauder basis is due to S. Banach. Banach spaces with bases are presented in a natural way as sequence spaces and are the simplest among all Banach spaces. Almost all results of this section are due to S. Banach [Ban]. They can also be found in the book of J. Lindenstrauss and L. Tzafriri [LT] or in the book of B. Beauzamy [Be3].

Definition I.1.1. Let X be a Banach space.

1. A sequence $(e_n)_{n\in\mathbb{N}}$ in X is a *basic sequence* if for all $x \in \overline{\text{Span}}[e_n,\ n \in \mathbb{N}]$ there exists a unique sequence $(a_n)_{n\in\mathbb{N}}$ of real numbers such that $x = \sum_{n=0}^{\infty} a_n e_n$.
2. $(e_n)_{n\in\mathbb{N}}$ is a *Schauder basis* of X if it is a basic sequence and if moreover $X = \overline{\text{Span}}[e_n,\ n \in \mathbb{N}]$.

Notation. If $(e_n)_{n\in\mathbb{N}}$ is a Schauder basis of X and $x \in X$, we denote by $x(k)$ the kth coordinate of x on $(e_n)_{n\in\mathbb{N}}$. We call the *support of x*

and denote by Supp(x) the subset of $\mathbb{N}$ such that $x(k)$ is nonzero. If Supp(x) is a finite subset of $\mathbb{N}$, we say that x is *finitely supported*.

Remark I.1.2. In this book, for infinite dimensional spaces we will not consider other type of bases beside Schauder bases. We shall therefore often omit the word Schauder. If X is finite dimensional, quantitative notions related to Schauder bases can be extended to algebraic bases and this will cause no confusion.

Proposition I.1.3. Let X be a Banach space and $(e_n)_{n\in\mathbb{N}}$ a sequence in X. Then $(e_n)_{n\in\mathbb{N}}$ is a basis of X if and only if three properties hold:

(i) $\forall n \in \mathbb{N}, \ e_n \neq 0$

(ii) There exists $K > 0$ such that for all finite sequences $(a_n)_{n\in\mathbb{N}}$ in $\mathbb{R}$ and all integers n, m with $n < m$, we have

$$\left\| \sum_{i=0}^{n} a_i e_i \right\| \leq K \left\| \sum_{i=0}^{m} a_i e_i \right\|$$

(iii) $X = \overline{\mathrm{Span}[e_n, \ n \in \mathbb{N}]}$

Proof. Suppose first that $(e_n)_{n\in\mathbb{N}}$ is a basis of X. Then (i) and (iii) are true by definition. Let us prove (ii):

$$\text{For } x = \sum_{i=0}^{\infty} a_i e_i \in X, \text{ set } |\,x\,| = \sup_n \left\| \sum_{i=0}^{n} a_i e_i \right\|$$

Then, since the sequence $(\sum_{i=0}^{n} a_i e_i)_{n\in\mathbb{N}}$ converges in X to x, we get

$$\| x \| < |\,x\,| < +\infty$$

$|\cdot|$ is obviously a norm on X. Let us prove that $(X, |\cdot|)$ is a Banach space: Let $\epsilon > 0$ and $(x^{(n)})_{n\in\mathbb{N}}$ be a Cauchy sequence in $(X, |\cdot|)$. Define the coordinates of $x^{(n)}$ by

$$x^{(n)} = \sum_{i=0}^{\infty} a_i^{(n)} e_i.$$

Then there exists $n_\epsilon \in \mathbb{N}$ such that

$$\forall p,q \geq n_\epsilon, \quad \sup_{n\in\mathbb{N}} \left\| \sum_{i=0}^{n} (a_i^{(p)} - a_i^{(q)}) e_i \right\| \leq \epsilon.$$

This implies that, for every fixed i, $(a_i^{(p)})_{p \in \mathbb{N}}$ is a real Cauchy sequence and thus converges. Let a_i be its limit, and define

$$x = \sum_{i=0}^{\infty} a_i e_i.$$

We have to prove that $x \in E$ and $\| x^{(n)} - x \| \xrightarrow[n \to +\infty]{} 0$. Since $x^{(n_\epsilon)}$ belongs to X, there exists $m_\epsilon \in \mathbb{N}$ such that

$$\forall N \geq m \leq m_\epsilon, \qquad \left\| \sum_{i=m}^{N} a_i^{(n_\epsilon)} e_i \right\| \leq \epsilon$$

We deduce from that, that for all $p \geq n_\epsilon$ and N, m such that $N \geq m \geq m_\epsilon$,

$$\left\| \sum_{i=m}^{N} a_i^{(p)} e_i \right\| \leq \left\| \sum_{i=m}^{N} (a_i^{(p)} - a_i^{(n_\epsilon)}) e_i \right\| + \left\| \sum_{i=m}^{N} a_i^{(n_\epsilon)} e_i \right\|$$

$$\leq \left\| \sum_{i=0}^{N} (a_i^{(p)} - a_i^{(n_\epsilon)}) e_i \right\| + \left\| \sum_{i=0}^{m} (a_i^{(p)} - a_i^{(n_\epsilon)}) e_i \right\|$$

$$+ \left\| \sum_{i=m}^{N} a_i^{(n_\epsilon)} e_i \right\|$$

$$\leq 3\epsilon$$

and thus for all $N \geq m \geq m_\epsilon$, $\| \sum_{i=m}^{N} a_i e_i \| \leq 3\epsilon$. Then $(\sum_{i=0}^{n} a_i e_i)_{n \in \mathbb{N}}$ is convergent in $(X, \|\cdot\|)$ and the vector $x = \sum_{i=0}^{\infty} a_i e_i$ belongs to X.

Moreover, for $p \geq n_\epsilon$, we get

$$\left\| \sum_{i=0}^{n} (a_i^{(p)} - a_i) \right\| \leq \left\| \sum_{i=0}^{m_\epsilon} (a_i^{(p)} - a_i) e_i \right\| + \left\| \sum_{i=m_\epsilon+1}^{n} a_i^{(p)} e_i \right\|$$

$$+ \left\| \sum_{i=m_\epsilon+1}^{n} a_i e_i \right\|$$

$$\leq \left\| \sum_{i=0}^{m_\epsilon} (a_i^{(p)} - a_i) e_i \right\| + 6\epsilon$$

Thus if $p \geq n_\epsilon$, then

$$\| x^{(p)} - x \| \leq \left\| \sum_{i=0}^{m_\epsilon} (a_i^{(p)} - a_i) e_i \right\| + 6\epsilon$$

The first term tends to 0 when $p \to +\infty$, and this proves that $(x^{(p)})_{p \in \mathbb{N}}$ converges to x in $(X, |\cdot|)$. Thus, $(X, |\cdot|)$ is a Banach space. Hence, since the identity from $(X, |\cdot|)$ into $(X, \|\cdot\|)$ is continuous and bijective between two Banach spaces, it is an isomorphism by the open mapping theorem, which implies that there exists $K > 0$ such that

$$\forall x \in X, \qquad |x| \leq K \|x\|$$

This proves (ii).

On the other hand, suppose that (i), (ii), and (iii) are realized. Let us first prove the *uniqueness of the decomposition*: Suppose that there exists $x \in X$ such that $x = \sum_{i=0}^{\infty} a_i e_i = \sum_{i=0}^{\infty} b_i e_i$. Then $|a_1 - b_1| \|e_1\| \leq K \|\sum_{i=0}^{\infty} (a_i - b_i)e_i\| = 0$, where K is given by (ii). Thus $a_1 = b_1$ by (i). We finish the proof by induction on the co-ordinates a_n and b_n.

Existence of the decomposition: Let

$$F = \left\{ x \in X / \exists (a_i)_{i \in \mathbb{N}} \in \mathbb{R}^{\mathbb{N}}, \, x = \sum_{i=0}^{\infty} a_i e_i \right\}$$

Then Span$[e_i, \, i \in \mathbb{N}]$ is a subspace of F, and by (iii) $\overline{F} = X$. Let us prove that $F = \overline{F}$: As above, define $x^{(n)} = \sum_{i=0}^{\infty} a_i^{(n)} e_i$ and suppose that $x^{(n)}{}_{n \in \mathbb{N}}$ converges to some x in X. Then $(a_i^{(n)})_{n \in \mathbb{N}}$ is a real Cauchy sequence for all $i \in \mathbb{N}$ by (i) and (ii), and then converges to some $a_i \in \mathbb{R}$. Then, for all $\epsilon > 0$, there exists $n_\epsilon \in \mathbb{N}$ such that, for all $p, q \geq n_\epsilon$,

$$\left\| \sum_{i=0}^{\infty} (a_i^{(p)} - a_i^{(q)})e_i \right\| \leq \epsilon$$

Thus, by (ii), for all $p, q \geq n_\epsilon$,

$$\underset{N}{\text{Sup}} \left\| \sum_{i=0}^{N} (a_i^{(p)} - a_i^{(q)})e_i \right\| \leq K \left\| \sum_{i=0}^{\infty} (a_i^{(p)} - a_i^{(q)})e_i \right\| \leq K\epsilon$$

Also, for all $p \geq n_\epsilon$,

$$\underset{N}{\text{Sup}} \left\| \sum_{i=0}^{N} (a_i - a_i^{(p)})e_i \right\| \leq K\epsilon$$

Moreover, there exists $m_\epsilon \in \mathbb{N}$ such that for all $N \geq m \geq m_\epsilon$,

$$\left\| \sum_{i=m}^{N} a^{(n_\epsilon)} e_i \right\| \leq \epsilon$$

Hence for all $N \geq m \geq m_\epsilon$,

$$\left\| \sum_{i=m}^{N} a_i e_i \right\| \leq \left\| \sum_{i=m}^{N} (a_i - a_i^{(n_\epsilon)}) e_i \right\| + \left\| \sum_{i=m}^{N} a_i^{(n_\epsilon)} e_i \right\| \leq 2K\epsilon + \epsilon$$

This proves that $(\sum_{i=0}^{n} a_i e_i)_{n \in \mathbb{N}}$ converges in X to some $x' \in X$.

It remains to prove that $x = x'$: By the same arguments as above, for all $\epsilon > 0$, there exist n_ϵ and N_ϵ such that, for all $p \geq n_\epsilon$,

$$\left\| \sum_{i=0}^{\infty} a_i^{(p)} e_i - x \right\| \leq \epsilon, \quad \left\| \sum_{i=N_\epsilon+1}^{\infty} a_i^{(p)} e_i \right\| \leq \epsilon, \quad \left\| \sum_{i=N_\epsilon+1}^{\infty} a_i e_i \right\| \leq \epsilon$$

We can choose $p_\epsilon \geq n_\epsilon$ such that $\| \sum_{i=0}^{N_\epsilon} a_i e_i - \sum_{i=0}^{N_\epsilon} a_i^{(p_\epsilon)} e_i \| \leq \epsilon$. Then we can write

$$\| x' - x \| = \left\| \sum_{i=0}^{\infty} a_i e_i - x \right\|$$

$$\leq \left\| \sum_{i=0}^{\infty} a_i e_i - \sum_{i=0}^{N_\epsilon} a_i e_i \right\| + \left\| \sum_{i=0}^{N_\epsilon} a_i e_i - \sum_{i=0}^{N_\epsilon} a_i^{(p_\epsilon)} e_i \right\|$$

$$+ \left\| \sum_{i=0}^{N_\epsilon} a_i^{(p_\epsilon)} e_i - \sum_{i=0}^{\infty} a_i^{(p_\epsilon)} e_i \right\| + \left\| \sum_{i=0}^{\infty} a_i^{(p_\epsilon)} e_i - x \right\|$$

$$\leq 4\epsilon$$

So $x = x'$ and $F = \overline{F} = X$, which concludes the proof of this proposition.

Remark I.1.4. Another way to express (ii) of Proposition I.1.3 is to say that the sequence of projections $(P_n)_{n \in \mathbb{N}}$ of X, defined by:

$$P_n \left(\sum_{i=0}^{\infty} a_i e_i \right) = \sum_{i=0}^{n} a_i e_i$$

has norms uniformly bounded by K.

Notation. We will always denote by $(P_n)_{n \in \mathbb{N}}$ the sequence of projections associated with a given sequence $(x_n)_{n \in \mathbb{N}}$ by $P_n(\sum_{i=0}^{\infty} a_i x_i) = \sum_{i=0}^{n} a_i x_i$.

Definition I.1.5

1. If X is a Banach space with a basis $(e_n)_{n \in \mathbb{N}}$, the *basis constant* of $(e_n)_{n \in \mathbb{N}}$ will be $\text{Sup}\{\| P_n \|, n \in \mathbb{N}\}$, where $(P_n)_{n \in \mathbb{N}}$ is the sequence of projections defined above.
2. If the basis constant of $(e_n)_{n \in \mathbb{N}}$ is 1, $(e_n)_{n \in \mathbb{N}}$ is called a *monotone basis*.
3. If $\| e_n \| = 1$ for all n, $(e_n)_{n \in \mathbb{N}}$ is called a *normalized basis*.

Examples I.1.6

1. $\ell_p (1 \le p < +\infty)$ and c_0 have monotone bases. The unit vectors $e_n = (0,0, \ldots 0,1,0, \ldots)$, where 1 is at the nth rank, form a monotone and normalized basis in each of these spaces.
2. The summing basis $(s_n)_{n \in \mathbb{N}}$ of c_0, defined by $s_n = \sum_{k=0}^{n} e_k$ for all $n \in \mathbb{N}$ is not monotone; indeed, if we take $x = \sum_{n \in \mathbb{N}}^{N} s_n - s_{N+1}$, then $\| x \| = N$ and $\| \sum_{n \in \mathbb{N}}^{N} s_n \| = N + 1$.
3. There is a natural monotone basis in $L_p[0,1]$ $(1 \le p < +\infty)$, called the Haar basis (resp. the normalized Haar basis).

    Let us define the Haar functions by

$$
\begin{aligned}
h_0(\omega) &= 1 & \omega &\in [0,1] \\
h_1(\omega) &= 1 & \omega &\in [0,\tfrac{1}{2}] \\
&= -1 & \omega &\in [\tfrac{1}{2},1] \\
h_2(\omega) &= 1 & \omega &\in [0,\tfrac{1}{4}[\\
&= -1 & \omega &\in [\tfrac{1}{4},\tfrac{1}{2}[\\
&= 0 & &\text{otherwise} \\
h_3(\omega) &= -1 & \omega &\in [\tfrac{1}{2},\tfrac{3}{4}[\\
&= -1 & \omega &\in [\tfrac{3}{4},1] \\
&= 0 & &\text{otherwise} \\
&\;\;\vdots
\end{aligned}
$$

For $k \in \mathbb{N}$, $1 \le l \le 2^k$ and $n = 2^k + l - 1$

$$
h_n(\omega) = 1 \qquad \omega \in \left[\frac{2l - 2}{2^{k+1}}, \frac{2l - 1}{2^{k+1}} \right[
$$

$$= -1 \qquad \omega \in \left[\frac{2l-1}{2^{k+1}}, \frac{2l}{2^{k+1}}\right[$$

$$= 0 \qquad \text{otherwise}$$

$$\vdots$$

Then $(h_n)_{n\in\mathbb{N}}$ [resp. $(h_n/\| h_n \|)_{n\in\mathbb{N}}$] is a monotone basis (resp. normalized monotone basis) of $L_p[0,1]$. Let us prove it for $(h_n)_{n\in\mathbb{N}}$; of course, (i) and (iii) of Proposition I.1.3 are true because the dyadic intervals generate the Borel σ-field in $[0,1]$. (ii) is also verified with $K = 1$ because if $f = \sum_{i=0}^{n} a_i h_i$ and $g = \sum_{i=0}^{n+1} a_i h_i$ the only difference between f and g occurs on the last dyadic interval where f is constant. Let us call this constant b. Then g takes two values, $b + a_{n+1}$ on the first half and $b - a_{n+1}$ on the second half of this interval. Since $p \geq 1$, the real function $| 1 + t |^p + | 1 - t |^p - 2$ is positive on $\mathbb{R}$, and this implies that

$$| b + a_{n+1} |^p + | b - a_{n+1} |^p \geq 2 | b |^p$$

This proves that

$$\| g \|_p \geq \| f \|_p$$

4. Nonseparable spaces cannot have a basis. This is the case for ℓ_∞ and $L_\infty[0,1]$.
5. We will see later (see Theorem I.5.1) that there exists a separable Banach space without any basis. The first example of such a space is due to P. Enflo [E1].

Moreover, the following result shows that even if a Banach space X does not have a basis, it contains a basic sequence.

Proposition I.1.7. Every infinite-dimensional Banach space X contains a basic sequence.

To prove Proposition I.1.7, we need a definition and a lemma.

Definition I.1.8. Let A be a relatively compact set in a Banach space Y. A finite set $\{y_0, \ldots, y_n\}$ in A such that for all $y \in A$ there exists $i \in \{0, \ldots, n\}$ such that $\| y - y_i \| \leq \epsilon$ will be called an ϵ-net in A.

Note that if A is a bounded set in a finite-dimensional space Y, it is relatively compact. Thus for all $\epsilon > 0$ there exists an ϵ-net in A.

Lemma I.1.9. Let $Y \subset X$, dim $Y < +\infty$, and $\epsilon > 0$. Then there exists $x \in X$, $\| x \| = 1$ such that for all $y \in Y$ and all $\lambda \in \mathbb{R}$, we have $\| y \| \leq (1 + \epsilon)\| y + \lambda x \|$.

Proof of Lemma I.1.9. We can suppose without loss of generality that $\epsilon < 1$. Let $(y_i)_{i=0,\ldots,n}$ be an $\epsilon/2$-net in S_Y. By the Hahn-Banach theorem, for all $i = 0, \ldots, n$ there exists $y_i^* \in X^*$ such that $y_i^*(y_i) = 1$. Since X is infinite dimensional, there exists $x \in X$ such that:

$$\text{For all } i = 0, \ldots, n, \qquad y_i^*(x) = 0$$

Now let $y \in Y$, $y \neq 0$, and $i_0 \in \{0, \ldots, n\}$ such that

$$\left\| \frac{y}{\| y \|} - y_{i_0} \right\| \leq \frac{\epsilon}{2}$$

Then

$$\left\| \frac{y}{\| y \|} + \lambda x \right\| \geq \| y_{i_0} + \lambda x \| - \frac{\epsilon}{2} \geq y_{i_0}^*(y_{i_0} + \lambda x) - \frac{\epsilon}{2} = 1 - \frac{\epsilon}{2}$$

So if we change λ to $\lambda/\| y \|$, we get $(1 - \epsilon/2) \| y \| \leq \| y + \lambda x \|$.

Proof of Proposition I.1.7. $\epsilon > 0$ being given, we choose a sequence of positive numbers $(\epsilon_n)_{n\in\mathbb{N}}$ such that

$$\prod_{n\in\mathbb{N}} (1 + \epsilon_n) \leq 1 + \epsilon$$

Also choose $x_0 \in X$ with $\| x_0 \| = 1$.

We build a sequence $(x_n)_{n\in\mathbb{N}}$ by induction according to Lemma I.1.9. Suppose that $\{x_0, \ldots, x_n\}$ have already been chosen. Then, if we take $Y = \text{Span}[x_i, 0 \leq i \leq n]$ and $\epsilon = \epsilon_n$, there exists x_{n+1} such that for all $(a_0, \ldots, a_{n+1}) \in \mathbb{R}^{n+2}$,

$$\left\| \sum_{i=0}^n a_i x_i \right\| \leq (1 + \epsilon_n) \left\| \sum_{i=0}^{n+1} a_i x_i \right\|$$

So $(x_n)_{n\in\mathbb{N}}$ will be a basic sequence with constant $K = \prod_{n=0}^{\infty} (1 + \epsilon_n) \leq 1 + \epsilon$.

The next result gives another way to find basic sequences in special situations.

Theorem I.1.10. Let $(x_n)_{n\in\mathbb{N}}$ be a sequence in a Banach space X which is not norm-convergent to 0.

1. If $(x_n)_{n\in\mathbb{N}}$ is $\sigma(X,X^*)$-convergent to 0, or
2. If $(x_n)_{n\in\mathbb{N}}$ is $\sigma(X,X^*)$-Cauchy and not $\sigma(X,X^*)$-convergent, then $(x_n)_{n\in\mathbb{N}}$ has a basic subsequence.

Proof. Let $(\epsilon_k)_{k\in\mathbb{N}}$ be a sequence of positive numbers such that $0 < \prod_{i=0}^{\infty}(1 - \epsilon_k) < +\infty$. Extracting a subsequence, if necessary, we can suppose that $\| x_n \| \geq \delta > 0$ for all n.

We proceed in both cases 1 and 2 by induction.

1. Suppose that $(x_n)_{n\in\mathbb{N}}$ is $\sigma(X,X^*)$-convergent to 0 and that $n_0, \ldots, n_k$ have already been chosen such that $n_0 = 0$ and for all $(a_i)_{0 \leq i \leq k}$ in $\mathbb{R}$ and all $l < k$,

$$\left\| \sum_{i=0}^{l} a_i x_{n_i} \right\| \leq (1 + \epsilon_k) \left\| \sum_{i=0}^{k} a_i x_{n_i} \right\|$$

Set $Y_k = \overline{\mathrm{Span}}\,[x_{n_0}, \ldots, x_{n_k}]$ and let $\{y_0, \ldots, y_{N_k}\}$ be an $(\epsilon_{k+1}/4)$-net in S_{Y_k}. By the Hahn-Banach theorem, we can choose $\{y_0^*, \ldots, y_{N_k}^*\}$ in S_{X^*} such that $y_i^*(y_i) \geq 1 - \epsilon_{k+1}/4$ for $i = 0, \ldots, N_k$.

By hypothesis there exists a rank n_{k+1} such that

$$| y_i^*(x_{n_{k+1}})| \leq \frac{\epsilon_{k+1}}{4} \quad \text{for } i = 0, \ldots, N_k$$

Then, to evaluate $\| \sum_{i=0}^{k+1} \alpha_i x_{n_i} \|$ for $\alpha_0 \cdots \alpha_{k+1} \in \mathbb{R}^{k+2}$, we consider two cases, depending on the value of $| \alpha_{k+1} |$:
First case:

$$| \alpha_{k+1} | \geq 2 \frac{\| \sum_{i=0}^{k} \alpha_i x_{n_i} \|}{\| x_{n_{k+1}} \|}$$

Then

$$\left\| \sum_{i=0}^{k+1} \alpha_i x_{n_i} \right\| \geq \| \alpha_{k+1} x_{n_{k+1}} \| - \left\| \sum_{i=0}^{k} \alpha_i x_{n_i} \right\| \geq \left\| \sum_{i=0}^{k} \alpha_i x_{n_i} \right\|$$

Second case:

$$| \alpha_{k+1} | < 2 \frac{\| \sum_{i=0}^{k} \alpha_i x_{n_i} \|}{\| x_{n_{k+1}} \|}$$

Let i_0 be such that

$$\left\| \frac{\sum_{i=0}^{k} \alpha_i x_{n_i}}{\| \sum_{i=0}^{k} \alpha_i x_{n_i} \|} - y_{i_0} \right\| \leq \frac{\epsilon_{k+1}}{4}$$

Then we can write

$$\left\| \sum_{i=0}^{k+1} \alpha_i x_{n_i} \right\| \geq y_{i_0}^* \left(\sum_{i=0}^{k} \alpha_i x_{n_i} + \alpha_{k+1} x_{n_{k+1}} \right)$$

$$\geq \left| y_{i_0}^* \left(y_{i_0} \left\| \sum_{i=0}^{k} \alpha_i x_{n_i} \right\| \right) \right|$$

$$- \left| y_{i_0}^* \left(y_{i_0} \left\| \sum_{i=0}^{k} \alpha_i x_{n_i} \right\| - \sum_{i=0}^{k} \alpha_i x_{n_i} \right) \right|$$

$$- | \alpha_{k+1} | \, | y_{i_0}^*(x_{n_{k+1}}) |$$

$$\geq \left\| \sum_{i=0}^{k} \alpha_i x_{n_i} \right\| \left[1 - \frac{\epsilon_{k+1}}{4} - \frac{\epsilon_{k+1}}{4} - \frac{2\epsilon_{k+1}}{4} \right]$$

$$\geq \left\| \sum_{i=0}^{k} \alpha_i x_{n_i} \right\| (1 - \epsilon_{k+1})$$

In both cases, we have $\| \sum_{i=0}^{k+1} \alpha_i x_{n_i} \| \geq \| \sum_{i=0}^{k} \alpha_i x_{n_i} \| (1 - \epsilon_{k+1})$. Going on inductively, we built this way a sequence $(x_{n_k})_{k \in \mathbb{N}}$ that will be basic with basis constant: $1/\prod_{i=0}^{\infty} (1 - \epsilon_k)$.

2. Suppose now that $(x_n)_{n \in \mathbb{N}}$ is $\sigma(X,X^*)$-Cauchy non-$\sigma(X,X^*)$-convergent. We need a definition and a lemma:

Definition. A set Z in X^* will be *norming for X* if there exists a constant $C > 0$ such that for all $x \in X$, $\text{Sup}_{z \in Z}\, z(x) \geq C \| x \|$.

By hypothesis there exists $x^{**} \in X^{**}$ such that $(x_n)_{n \in \mathbb{N}}$ is $\sigma(X^{**},X^*)$-convergent to x^{**}. Let $Z = \text{Ker } x^{**} = \{z \in X^*, x^{**}(z) = 0\} \subseteq X^*$.

Lemma I.1.11. Z is norming for X.

Proof of Lemma I.1.11. Let $g \in X^*$, $\| g \| = 1$ be such that $X^* = Z \oplus \mathbb{R}g$ and suppose that Z is not norming. Then there exists $(x_n)_{n \in \mathbb{N}} \subset X$, $\| x_n \| = 1$ such that the sequence $(\mathrm{Sup}_{z \in B_Z} z(x_n))_{n \in \mathbb{N}}$ converges to 0.

Since $(g(x_n))_{n \in \mathbb{N}}$ is a bounded sequence in $\mathbb{R}$, passing to a subsequence, we can suppose that $g(x_n)$ converges to some $a \in \mathbb{R}$. We consider two cases depending on the value of a:

*First case: a $= 0$. Then

$$\| x_n \| = \mathrm{Sup}\{(z + \alpha g)(x_n)/z \in Z, \alpha \in \mathbb{R}, \| z + \alpha g \| \leq 1\}$$
$$= \mathrm{Sup}\{z(x_n) + \alpha g(x_n)/z \in Z, \alpha \in \mathbb{R}, \| z + \alpha g \| \leq 1\}$$
$$\xrightarrow[n \to +\infty]{} 0$$

This is a contradiction.

*Second case: a $\neq 0$. Then

$$\| x_n - ax^{**} \| = \mathrm{Sup}\{z(x_n) + \alpha g(x_n - ax^{**})/z \in Z, \alpha \in \mathbb{R},$$
$$\| z + \alpha g \| = \leq 1\} \xrightarrow[n \to +\infty]{} 0$$

So $(x_n)_{n \in \mathbb{N}}$ converges to x^{**} in norm and $x^{**} \in X$, which is a contradiction again. Thus Z is norming for X in both cases.

We finish the proof of Theorem I.1.10 by the same techniques as in the first situation, replacing X^* by Z and $\| x \|$ by $| x | = \mathrm{Sup}_{z \in B_Z} z(x)$. Indeed, since these two norms are equivalent, any sequence that is basic for the first norm is also basic for the other.

The two next results describe situations in which we know that two basic sequences behave in the same way.

Definition I.1.12. Two basic sequences $(x_n)_{n \in \mathbb{N}}$ and $(y_n)_{n \in \mathbb{N}}$ are *eqivalent* if the convergence of a series $\sum_{n=0}^{\infty} a_n x_n$ implies that of the series $\sum_{n=0}^{\infty} a_n y_n$ and conversely. We will denote that property by $(x_n)_{n \in \mathbb{N}} \sim (y_n)_{n \in \mathbb{N}}$.

Proposition I.1.13.   Let $(x_n)_{n \in \mathbb{N}}$ and $(y_n)_{n \in \mathbb{N}}$ be two basic sequences in a Banach space. Then the following properties are equivalent

   (i) $(x_n)_{n \in \mathbb{N}} \sim (y_n)_{n \in \mathbb{N}}$
   (ii) There exists $C > 0$ such that for all finite sequences of real numbers $(a_i)_{i \in \mathbb{N}}$,

$$ 1/C \left\| \sum_{i=0}^{\infty} a_i y_i \right\| \leq \left\| \sum_{i=0}^{\infty} a_i x_i \right\| \leq C \left\| \sum_{i=0}^{\infty} a_i y_i \right\| $$

   (iii) $\overline{\mathrm{Span}}[x_n, n \in \mathbb{N}]$ and $\overline{\mathrm{Span}}[y_n, n \in \mathbb{N}]$ are isomorphic.

Proof.   (iii) $\Leftrightarrow$ (ii) $\Rightarrow$ (i) are obvious. Let us prove (i) $\Rightarrow$ (ii).
   Let T_n be defined by

$$ T_n: \overline{\mathrm{Span}}[x_n, n \in \mathbb{N}] \rightarrow \overline{\mathrm{Span}}[y_n, \in \mathbb{N}] $$

$$ x = \sum_{i=0}^{\infty} a_i x_i \mapsto T_n x = \sum_{i=0}^{n} a_i y_i $$

Then there exists a constant $C(n) > 0$ (depending on n) such that for all $(a_i)_{i=0,\dots,n}$ in $\mathbb{R}^{n+1}$,

$$ \left\| \sum_{i=0}^{n} a_i y_i \right\| \leq C(n) \left\| \sum_{i=0}^{n} a_i x_i \right\| \leq C(n) K \left\| \sum_{i=0}^{\infty} a_i x_i \right\| $$

where K is the basis constant of $(x_n)_{n \in \mathbb{N}}$. This proves that T_n is continuous with $\| T_n \| \leq C(n)K$.
   Moreover, because of hypothesis (i), for a fixed $x = \sum_{i=0}^{\infty} a_i x_i$, we have

$$ \operatorname*{Sup}_{n \in \mathbb{N}} \| T_n x \| = \operatorname*{Sup}_{n \in \mathbb{N}} \left\| \sum_{i=0}^{n} a_i y_i \right\| < +\infty $$

Then we can apply the Banach-Steinhaus theorem, which says that

$$ \operatorname*{Sup}_{n \in \mathbb{N}} \| T_n \| = C < +\infty $$

That means that the right-hand side inequality of (ii) holds with this constant C. The left-hand side inequality is proved in an obvious analogous way.

Definition I.1.14.   If $(x_n)_{n \in \mathbb{N}}$ and $(y_n)_{n \in \mathbb{N}}$ satisfy

$$1/C \left\| \sum_{i=0}^{\infty} a_i y_i \right\| \leq \left\| \sum_{i=0}^{\infty} a_i x_i \right\| \leq C \left\| \sum_{i=0}^{\infty} a_i y_i \right\|$$

for all $(a_i)_{i \in \mathbb{N}} \subset \mathbb{R}^{\mathbb{N}}$, we will say that $(x_n)_{n \in \mathbb{N}}$ and $(y_n)_{n \in \mathbb{N}}$ are *C-equivalent* and we will write

$$(x_n)_{n \in \mathbb{N}} \overset{C}{\sim} (y_n)_{n \in \mathbb{N}}$$

Remark.   If $(x_n)_{n \in \mathbb{N}}$ and $(y_n)_{n \in \mathbb{N}}$ are *C*-equivalent, then $\overline{\text{Span}}[x_n, n \in \mathbb{N}]$ and $\overline{\text{Span}}[y_n, n \in \mathbb{N}]$ are *C*-isomorphic.

Proposition I.1.15.   Let $(x_n)_{n \in \mathbb{N}}$ be a basic sequence in a Banach space X that is not convergent to 0 in X, and let $(y_n)_{n \in \mathbb{N}}$ be a sequence in X such that

$$\sum_{n=0}^{\infty} \| x_n - y_n \| < +\infty$$

Then there exist a subsequence $(x_n')_{n \in \mathbb{N}}$ of $(x_n)_{n \in \mathbb{N}}$ and a subsequence $(y_n')_{n \in \mathbb{N}}$ of $(y_n)_{n \in \mathbb{N}}$ such that $(x_n')_{n \in \mathbb{N}} \sim (y_n')_{n \in \mathbb{N}}$.

Proof.   By extracting appropriate subsequences, we can suppose that

$$\begin{cases} \forall n \in \mathbb{N}, \quad \| x_n \| \geq \delta > 0 \\ \sum_{i=0}^{\infty} \| x_n - y_n \| < \dfrac{\delta}{2K} \end{cases}$$

(where K is the basis constant of $(x_n)_{n \in \mathbb{N}}$).

For $x \in X$, $x = \sum_{i=0}^{\infty} a_i x_i$, set $Tx = \sum_{i=0}^{\infty} a_i y_i$. Then

$$\| x - Tx \| \leq \sum_{i=0}^{\infty} | a_n | \, \| x_n - y_n \| \leq \underset{n \in \mathbb{N}}{\text{Max}} |a_n| \sum_{i=0}^{\infty} \| x_n - y_n \|$$

$$\leq \frac{2K \| x \|}{\delta} \sum_{i=0}^{\infty} \| x_n - y_n \| < \| x \|$$

Thus $\| I - T \| < 1$ and T is an isomorphism from $\overline{\text{Span}}[x_n, n \in \mathbb{N}]$ onto $\overline{\text{Span}}[y_n, n \in \mathbb{N}]$. This proves Proposition I.1.15 by (iii) of Proposition I.1.13.

The notion of *blocks* on a sequence is very useful and has many applications in this book. Let us give the definition.

Definition I.1.16

1. Let $(x_n)_{n\in\mathbb{N}}$ be a sequence in a Banach space X. A sequence $(u_n)_{n\in\mathbb{N}}$ is a *sequence of blocks* of $(x_n)_{n\in\mathbb{N}}$ if there exist a nondecreasing sequence $(p_j)_{j\in\mathbb{N}}$ in $\mathbb{N}$ and a sequence $(a_n)_{n\in\mathbb{N}} \subset \mathbb{R}^\mathbb{N}$ such that

$$\forall j \in \mathbb{N}, \qquad u_j = \sum_{n=p_j+1}^{p_j+1} a_n x_n$$

2. If $(x_n)_{n\in\mathbb{N}}$ is a basic sequence, then such a sequence of blocks $(u_n)_{n\in\mathbb{N}}$ is called a *block-basis* of $(x_n)_{n\in\mathbb{N}}$.
    (Note that it is clear that $(u_n)_{n\in\mathbb{N}}$ is basic with a constant less than the basis constant of $(x_n)_{n\in\mathbb{N}}$).

The following technical result will have two corollaries, on the behavior of weakly null sequences in a Banach space with a basis.

Proposition I.1.17.   Let $(x_n)_{n\in\mathbb{N}}$ be a basic sequence in a Banach space X and suppose that $(y_k)_{k\in\mathbb{N}}$ is a sequence in $\overline{\text{Span}}[x_n, n \in \mathbb{N}]$. For all $k \in \mathbb{N}$, we can write the decomposition of y_k as $y_k = \sum_{n=0}^{\infty} a_k(n)x_n$. If we assume that $(y_k)_{k\in\mathbb{N}}$ does not converge to 0 and that $\lim_{k\to+\infty} a_k(n) = 0$ for all $n \in \mathbb{N}$, then there exists a subsequence $(y_{n_k})_{k\in\mathbb{N}}$ of $(y_k)_{k\in\mathbb{N}}$ that is equivalent to a block-basis of $(x_n)_{n\in\mathbb{N}}$.

Proof.   Fix $\epsilon > 0$. By extracting a subsequence, we can suppose that $\| y_k \| \geq \delta > 0$ for all $k \in \mathbb{N}$. Suppose that $y_{n_0}, \ldots, y_{n_k}$ have already been chosen such that there exist blocks $u_0, \ldots, u_k$ on $(x_n)_{n\in\mathbb{N}}$ (see Definition I.1.16) such that $\| y_{n_i} - u_i \| \leq \epsilon/2^i$ for all $i = 0, \ldots, k$. Let P_{k+1} be the last integer in the decomposition of u_k on $(x_n)_{n\in\mathbb{N}}$.
    By hypothesis, there exists $n_{k+1} \in \mathbb{N}$ such that

$$\left\| \sum_{n=0}^{P_{k+1}} a_{n_{k+1}}(n)x_n \right\| \leq \frac{\epsilon}{2^{k+2}}$$

Then there exists $P_{k+2} \in \mathbb{N}$ such that

$$\left\| \sum_{n=P_{k+2}+1}^{\infty} a_{n_{k+1}}(n)x_n \right\| \leq \frac{\epsilon}{2^{k+2}}$$

Take

$$u_{k+1} = \sum_{n=P_{k+1}+1}^{P_{k+2}} a_{n_{k+1}}(n)x_n$$

Then $\| y_{n_{k+1}} - u_{k+1} \| \le \epsilon/2^{k+1}$. So continuing inductively we can build a subsequence $(y_{n_k})_{k\in\mathbb{N}}$ and a block-basis $(u_k)_{k\in\mathbb{N}}$ verifying the hypothesis of Proposition I.1.15. This concludes Proposition I.1.17.

Corollary I.1.18. Let X be a Banach space with a basis $(e_n)_{n\in\mathbb{N}}$. Then if $(x_k)_{k\in\mathbb{N}}$ is a sequence in X, $\sigma(X,X^*)$-converging to 0 and not converging in norm, $(x_k)_{k\in\mathbb{N}}$ has a subsequence that is equivalent to a block-basis of $(e_n)_{n\in\mathbb{N}}$.

Proof. Obviously, if $(x_k)_{k\in\mathbb{N}}$ tends to 0 for $\sigma(X,X^*)$, for each n, the nth coordinate $x_k(n)$ of $(x_k)_{k\in\mathbb{N}}$ on $(e_n)_{n\in\mathbb{N}}$ tends to 0 when k tends to $+\infty$. We can apply Proposition I.1.17.

Corollary I.1.19. Let X be a Banach space with a basis $(e_n)_{n\in\mathbb{N}}$. If X contains a subspace Y, isomorphic to $\ell_p(1 \le p < +\infty)$ or c_0, there exists a block-basis of $(e_n)_{n\in\mathbb{N}}$ that is equivalent to the unit vector basis of ℓ_p or c_0.

Proof. We have to consider two different cases:

First case: If Y is isomorphic to ℓ_p $(1 < p < \infty)$ or c_0, since the unit vector basis of these spaces $\sigma(Y,Y^*)$-converges to 0, this result is a direct consequence of Corollary I.1.18 and the obvious fact that all subsequences of the unit vector basis of $\ell_p(1 \le p < +\infty)$ or c_0 are equivalent to the whole sequence.

Second case: Suppose there exists $(f_n)_{n\in\mathbb{N}} \subset X$ such that $(f_n)_{n\in\mathbb{N}}$ is equivalent to the unit vector basis of ℓ_1. Recall that $x(k)$ denotes the kth coordinate of an element $x \in X$ on the basis $(e_k)_{k\in\mathbb{N}}$.

We need a technical lemma that expresses that there always exists an element g in X, belonging to $\overline{\text{Span}}[f_n, n \ge N]$, and which has small first coordinates on $(e_n)_{n\in\mathbb{N}}$.

Lemma I.1.20. For every $k \in \mathbb{N}$, $\epsilon > 0$, and $N \in \mathbb{N}$, there exists g in $\overline{\text{Span}}[f_n, n \ge N]$ such that $\| g \| = 1$ and for all $j \le k$, $| g(j)| < \epsilon$.

Proof of Lemma I.1.20. If not, there exist $k \in \mathbb{N}$, $\epsilon > 0$, and $N \in \mathbb{N}$ such that for all g in $\overline{\mathrm{Span}}[f_n, n \geq N]$ with $\| g \| = 1$, there exists $j_0 \leq k$ such that $| g(j_0)| \geq \epsilon$. Then the linear map

$$\overline{\mathrm{Span}}[f_n, n \geq N] \to \ell_\infty^k$$

$$g \mapsto (g(0), g(1), \ldots, g(k))$$

is an isomorphism. Indeed, If K is the basis constant of $(e_n)_{n \in \mathbb{N}}$ we get

$$\epsilon \| g \| \leq \mathrm{Sup}\{| g(0)|, \ldots, | g(k)|\} \leq \frac{2K \| g \|}{\mathrm{Sup}\{\| e_n \|, n \in \mathbb{N}\}}$$

Of course, this is impossible and Lemma I.1.20 is proved.

This lemma proves Corollary I.1.19. Indeed, we are going to build by induction a sequence of blocks $(g_n)_{n \in \mathbb{N}}$ on $(e_n)_{n \in \mathbb{N}}$ that is equivalent to the unit vector basis of ℓ_1. In Lemma I.1.20, take $k = 0$, $\epsilon = \frac{1}{4}$, $N = 1$. Then there exists $g_0 \in \overline{\mathrm{Span}}[f_n, n > 0]$ such that

$$\begin{cases} | g_0(0)| < \frac{1}{4} \\ \| g_0 \| = 1 \end{cases}$$

Changing g_0 if necessary, we can suppose that g_0 has finite support on $(f_n)_{n \in \mathbb{N}}$ and that $\| g_0 \| \geq 1 - \frac{1}{2}$. Let N_1 be the last index in its support.

Take $k = 1$, $\epsilon = \frac{1}{8}$, $N = N_1$. There exists $g_1 \in \overline{\mathrm{Span}}[f_n, n > N_1]$ such that

$$\begin{cases} | g_1(j)| < \frac{1}{8} & \text{if } j = 0, 1 \\ \| g_1 \| = 1 \end{cases}$$

Again changing g_1 if necessary, we can suppose that g_1 has finite support on $(f_n)_{n \in \mathbb{N}}$ and that $\| g_1 \| \geq 1 - 1/2^2$. Let N_2 be the last index in its support. Take $k = 2$, $\epsilon = 1/2^4$, $N = N_2$. There exists $g_2 \in \overline{\mathrm{Span}}[f_n, n > N_2]$ such that

$$\begin{cases} | g_2(j)| < \dfrac{1}{2^4} & \text{if } j = 0, 1, 2 \\ \| g_2 \| = 1 \end{cases}$$

Continuing this construction, we get a sequence $(g_n)_{n \in \mathbb{N}}$ in X such that

$$\begin{cases} \| g_n \| \to 1 & \text{when } n \to +\infty \\ (g_n)_{n \in \mathbb{N}} & \text{are disjoint blocks on } (f_n)_{n \in \mathbb{N}} \\ g_n(j) \to 0 & \text{when } n \to +\infty \text{ for all } j \in \mathbb{N} \end{cases}$$

The two first properties imply that $(g_n)_{n \in \mathbb{N}}$ is equivalent to the unit vector basis of ℓ_1, and the third implies with Proposition I.1.17 that $(g_n)_{n \in \mathbb{N}}$ has a subsequence that is equivalent to a block basis of $(e_n)_{n \in \mathbb{N}}$.

2. BASES AND DUALITY

Even if X is a Banach space with a basis, in general its dual space X^* fails to have one. In this section, we give some results on the relations between the space and its dual space in terms of bases. These results can also be found in [LT] and partly in [Be3].

Definition I.2.1. A system $(e_n, e_n^*)_{n \in \mathbb{N}}$ is called a *biorthogonal system* if $(e_n)_{n \in \mathbb{N}}$ is a basis in a Banach space X and $e_n^* \in X^*$ is defined for all $n \in \mathbb{N}$ by

$$e_n^* \left(\sum_{i=0}^{\infty} a_i e_i \right) = a_n$$

The elements $(e_n^*)_{n \in \mathbb{N}}$ are called the *biorthogonal functionals*.

Proposition I.2.2. Let $(e_n, e_n^*)_{n \in \mathbb{N}}$ be a biorthogonal system and K be the basis constant of $(e_n)_{n \in \mathbb{N}}$. Then

1. $\forall n \in \mathbb{N}, \| e_n^* \| \leq 2K/\|e_n\|$.
2. $(e_n^*)_{n \in \mathbb{N}}$ is a basic sequence with constant less than K.

Proof. 1. For $x = \sum_{i=0}^{\infty} x(i)e_i$, we can write

$$| e_n^*(x) | = | x(n) | = \frac{\| x(n)e_n \|}{\| e_n \|} \leq \frac{2K \| x \|}{\| e_n \|}$$

which proves that $\| e_n^* \| \leq 2K/\| e_n \|$. To prove 2, let $(P_n)_{n \in \mathbb{N}}$ be the sequence of projections of X defined by

$$\forall x = \sum_{i=0}^{\infty} a_i e_i \in X \qquad P_n(x) = \sum_{i=0}^{n} a_i e_i \quad \text{(see Remark I.1.4)}$$

Then $\forall n \in \mathbb{N}, \| P_n \| \le K$.

Recall that the *transpose* T^* of an operator T from X to Y is an operator from Y^* to X^* defined by $T^*(y^*)(x) = y^*(x)$. If P_n^* is the transpose of P_n, we get, for $m \ge n$,

$$\forall x \in X \qquad P_n^* \left(\sum_{i=0}^{m} a_i e_i^* \right) (x) = \sum_{i=0}^{m} a_i e_i^*(P_n x) = \sum_{i=0}^{n} a_i e_i^*(x)$$

Since the sequence of projections (P_n^*) in X^* is uniformly bounded by K, this proves that $(e_n^*)_{n \in \mathbb{N}}$ is a basic sequence with basis constant less than K.

Let us give two definitions to make this situation more precise.

Definition I.2.3

1. A basis $(e_n)_{n \in \mathbb{N}}$ in a Banach space X is called *shrinking* if $(e_n^*)_{n \in \mathbb{N}}$ is a basis of X^*.
2. $(e_n)_{n \in \mathbb{N}}$ is said to be *boundedly complete* if for all $(a_n)_{n \in \mathbb{N}} \subset \mathbb{R}^{\mathbb{N}}$ such that $\text{Sup}_{n \in \mathbb{N}} \| \sum_{i=0}^{n} a_i e_i \| < +\infty$, the series $\sum_{i=0}^{\infty} a_i e_i$ converges in X.

Before giving the relation between these two definitions, let us prove a very useful characterization of shrinking bases.

Proposition I.2.4. Let $(e_n)_{n \in \mathbb{N}}$ be a basis of the Banach space X. Then $(e_n)_{n \in \mathbb{N}}$ is shrinking if and only if, for all $x^* \in X^*$, $(\| x^* |_{\overline{\text{Span}}[e_i, \, i \ge n]} \|)_{n \in \mathbb{N}}$ converges to 0 when $n \to +\infty$, where $x^* |_{\overline{\text{Span}}[e_i, \, i \ge n]}$ denotes the restriction of the linear functional x^* from X to $\overline{\text{Span}}[e_i, \, i \ge n]$.

Proof. If $(e_n^*)_{n \in \mathbb{N}}$ is a basis of X^*, then for all $x^* \in X^*$ we have

$$\| P_n^* x^* - x^* \| \xrightarrow[n \to +\infty]{} 0$$

where $(P_n^*)_{n \in \mathbb{N}}$ are the transposes of the projections $(P_n)_{n \in \mathbb{N}}$ associated with $(e_n)_{n \in \mathbb{N}}$ (see the proof of Proposition I.7.9).

Since $(P_{n-1}^* x^*)|_{\overline{\text{Span}[e_i,\ i \geq n]}} = 0$, we deduce that

$$\| x^* |_{\overline{\text{Span}[e_i,\ i \geq n]}} \| \xrightarrow[n \to +\infty]{} 0$$

On the other hand, suppose that

$$\| x^* |_{\overline{\text{Span}[e_i,\ i \geq n]}} \| \xrightarrow[n \to +\infty]{} 0$$

Let $x \in X$, $\| x \| = 1$, be given. Then

$$
\begin{aligned}
|(x^* - P_n^* x^*)(x)| &= | x^*(I - P_n)(x)| \\
&\leq \| x^* |_{\overline{\text{Span}[e_i,\ i \geq n]}} \| \| x \| \| I - P_n \| \\
&\leq (K + 1)\| x \| \| x^* |_{\overline{\text{Span}[e_i,\ i \geq n]}} \|
\end{aligned}
$$

where K is the basis constant of $(e_n)_{n \in \mathbb{N}}$.

The hypothesis implies that $\| x^* - P_n x^* \|$ tends to zero and thus that $(e_n^*)_{n \in \mathbb{N}}$ is a basis of X^*, so $(e_n)_{n \in \mathbb{N}}$ is shrinking.

We can now show the relation between shrinking bases and boundedly complete bases.

Proposition I.2.5. If $(e_n)_{n \in \mathbb{N}}$ is a shrinking basis of X, then $(e_n^*)_{n \in \mathbb{N}}$ is a boundedly complete basis of X^*.

Proof. If $\text{Sup}_n \| \sum_{i=0}^n a_i e_i^* \|$ is finite, then the sequence $u_n^* = \sum_{i=0}^n a_i e_i^*$ has a $\sigma(X^*,X)$-convergent subsequence $(u_{n_k}^*)_{k \in \mathbb{N}}$. Call u^* its $\sigma(X^*,X)$-limit. Then we have $u^*(e_n) = \lim_k (\sum_{i=0}^{n_k} a_i e_i^*)(e_n) = a_n$. Thus $u^* = \sum_{i=0}^\infty a_i e_i^*$ and the sequence $(\sum_{i=0}^n a_i e_i^*)_{n \in \mathbb{N}}$ converges to u^* by the uniqueness of the $\sigma(X^*,X)$-limit point. Since $(e_n)_{n \in \mathbb{N}}$ is shrinking, by Proposition I.2.6 , u^* is also the limit of $(u_n^*)_{n \in \mathbb{N}}$ in norm and this proves our assumption.

The next result gives the main property of boundedly complete bases.

Proposition I.2.6. If $(e_n)_{n \in \mathbb{N}}$ is a boundedly complete basis of X, then X is isomorphic to a dual space.

Proof. If $(e_n)_{n \in \mathbb{N}}$ is boundedly complete, with basis constant K, set

$$Z = \overline{\text{Span}}[e_n^*, n \in \mathbb{N}]$$

$$J : X \to Z^*$$

$$x \mapsto Jx, \qquad Jx(z) = z(x), \forall z \in Z$$

Then we are going to prove that J is an onto isomorphism. Indeed, if $x \in \overline{\text{Span}}[e_i, i \leq n]$, there exists $x^* \in X^*$ such that $x^*(x) = \|x\|$ and $\|x^*\| = 1$. As in Proposition I.2.2, denote by P_n^* the transposes of the projections P_n associated with $(e_n)_{n \in \mathbb{N}}$. Thus

$$
\begin{cases}
P_n^* x^*(x) = x^*(P_n x) = x^*(x) \\
P_n^* x^* = \sum_{i=0}^{n} x^*(e_i) e_i^* \in Z \\
\|P_n^* x^*\| \leq K \|x^*\| = K
\end{cases}
$$

and this implies, taking $z = P_n^* x^*$, that $\|Jx\| \geq \|x\|/K$. Since we also have by definition $\|Jx\| \leq \|x\|$, this proves by density that J is an isomorphism on its range.

The mapping J is surjective. Let $z^* \in Z^*$. Then for all $z \in Z$, we have

$$
\left\| \left(\sum_{i=0}^{n} z^*(e_i^*) Je_i \right) (z) \right\| = \left| \sum_{i=0}^{n} z^*(e_i^*) z(e_i) \right| = \left| z^* \left(\sum_{i=0}^{n} z(e_i) e_i^* \right) \right|
$$

$$
\leq \|z^*\| \left\| \sum_{i=0}^{n} z(e_i) e_i^* \right\| \leq K \|z^*\| \|z\|
$$

The sequence $(\sum_{i=0}^{n} z^*(e_i^*) Je_i)_{n \in \mathbb{N}}$ is uniformly bounded by $K \|z^*\|$. This implies that the sequence $(\sum_{i=0}^{n} z^*(e_i^*) e_i)_{n \in \mathbb{N}}$ is also uniformly bounded, and since $(e_i)_{i \in \mathbb{N}}$ is boundedly complete, it converges to $x = \sum_{i=0}^{\infty} z^*(e_i^*) e_i$. Clearly, $z^* = Jx$ and J is surjective. This proves that X is isomorphic to a dual space Z^*.

Examples I.2.7

1. If X has a shrinking basis, then X^* has a basis too and so is separable. Thus the unit vector basis of ℓ_1, the Haar basis of $L_1[0,1]$ (see I.1.6), is not shrinking.
2. If X is reflexive and has a basis $(e_n)_{n \in \mathbb{N}}$, then $(e_n)_{n \in \mathbb{N}}$ is shrinking; if not, then X^* is not equal to $\overline{\text{Span}}[e_n^*, n \in \mathbb{N}]$, and by the Hahn-Banach theorem, there exists $\xi \in X^{**}$ such that

$$
\xi \neq 0 \quad \text{and} \quad \xi(\overline{\text{Span}}[e_n^*, n \in \mathbb{N}]) = 0
$$

This is a contradiction because, since $X^{**} = X$, ξ can be written as

$$\xi = \sum_{i=0}^{\infty} a_i e_i$$

But $\xi(e_i^*) = a_i$, for all $i \in \mathbb{N}$ implies $a_i = 0$ for all $i \in \mathbb{N}$, and thus $\xi = 0$. $(e_n)_{n \in \mathbb{N}}$ is also obviously boundedly complete because of Proposition I.2.5.

3. The unit vector basis of c_0 is shrinking and not boundedly complete.
4. The unit vector basis of ℓ_1 is boundedly complete and not shrinking.

By combining the two notions of shrinking basis and boundedly complete basis, we get the following remarkable characterization of reflexivity, due to R. C. James [J3].

Theorem I.2.8. Let $(e_n)_{n \in \mathbb{N}}$ be a basis of a Banach space X. Then X is reflexive if and only if $(e_n)_{n \in \mathbb{N}}$ is shrinking and boundedly complete.

Proof. We already noted that if X is reflexive, then $(e_n)_{n \in \mathbb{N}}$ is shrinking and boundedly complete (see Examples I.2.7). Suppose now that $(e_n)_{n \in \mathbb{N}}$ is shrinking and boundedly complete and let $x^{**} \in X^{**}$. Then for all $y^* \in X^*$, since $(e_n)_{n \in \mathbb{N}}$ is shrinking, y^* has a decomposition on $(e_n^*)_{n \in \mathbb{N}}$; that is, $y^* = \sum_{i=0}^{\infty} a_i e_i^*$.

So we get, if $(P_n)_{n \in \mathbb{N}}$ is the sequence of projections associated with $(e_n)_{n \in \mathbb{N}}$,

$$\begin{cases} (P_n^{**} x^{**})(y^*) = x^{**}(P_n^* y^*) = x^{**} \left(\sum_{i=0}^{n} a_i e_i^* \right) \\ y^* \left(\sum_{i=0}^{n} x^{**}(e_i^*) e_i \right) = \sum_{i=0}^{n} x^{**}(e_i) y^*(e_i) = x^{**} \left(\sum_{i=0}^{n} a_i e_i^* \right) \end{cases}$$

Thus $P_n^{**} x^{**} = \sum_{i=0}^{n} x^{**}(e_i^*) e_i \in X$ and $\| P_n^{**} x^{**} \| \leq K \| x^{**} \|$ (where K is the basis constant of $(e_n)_{n \in \mathbb{N}}$). Then, since $(e_n)_{n \in \mathbb{N}}$ is boundedly complete, $P_n^{**} x^{**}$ converges in X to $\sum_{i=0}^{\infty} x^{**}(e_i^*) e_i$. Since $\sum_{i=0}^{\infty} x^{**}(e_i^*) e_i$ and x^{**} coincide on a dense subset of X^*, we have $x^{**} = \sum_{i=0}^{\infty} x^{**}(e_i^*) e_i \in X$, so $X = X^{**}$ and X is reflexive.

3. UNCONDITIONAL BASES

The structure of a space with a basis is simple because it is described
as a sequence space. However, in general not a lot more can be said.
A very useful class of bases with more precise properties is the class
of unconditional bases. The results of this part can also be found in
[LT] and [Be3].

Definition I.3.1.   A basic sequence $(e_n)_{n\in\mathbb{N}}$ is *unconditional* if any
convergent series $\sum_{i=0}^{\infty} a_i e_i$ *converges unconditionally*; that is, the series $\sum_{i=0}^{\infty} a_{\pi(i)} e_{\pi(i)}$ converges to the same limit for all permutations π
in $\mathbb{N}$.

Theorem I.3.2.   Let $(e_n)_{n\in\mathbb{N}}$ be a basic sequence in a Banach space
X. Then the following properties are equivalent:

(a) $(e_n)_{n\in\mathbb{N}}$ is unconditional.
(b)   If a series $\sum_{n=0}^{\infty} a_n e_n$ converges, then for all $\epsilon_n = \pm 1$, the
series $\sum_{n=0}^{\infty} \epsilon_n a_n e_n$ converges.
(b')   There exits $K_1 > 0$ such that, for all $\epsilon_n = \pm 1$ and all $(a_n)_{n\in\mathbb{N}}$
in $\mathbb{R}^{(\mathbb{N})}$,

$$\left\| \sum_{n=0}^{\infty} \epsilon_n a_n e_n \right\| \leq K_1 \left\| \sum_{n=0}^{\infty} a_n e_n \right\|$$

(c)   If a series $\sum_{n=0}^{\infty} a_n e_n$ converges, then for all $A \subset \mathbb{N}$, the series
$\sum_{n\in A} a_n e_n$ converges.
(c')   There exists $K_2 > 0$ such that, for all $A \subset \mathbb{N}$ and all $(a_n)_{n\in\mathbb{N}}$
in $\mathbb{R}^{(\mathbb{N})}$,

$$\left\| \sum_{n\in A} a_n e_n \right\| \leq K_2 \left\| \sum_{n=0}^{\infty} a_n e_n \right\|$$

(d)   If a series $\sum_{n=0}^{\infty} a_n e_n$ converges, then for all $(b_n)_{n\in\mathbb{N}} \in \mathbb{R}^{\mathbb{N}}$ such
that $| b_n | \leq | a_n |$ for all n, the series $\sum_{n=0}^{\infty} b_n e_n$ converges.
(d')   There exists $K_3 > 0$ such that, for all $(a_n)_{n\in\mathbb{N}}$ and $(b_n)_{n\in\mathbb{N}}$
in $\mathbb{R}^{(\mathbb{N})}$ with $| b_n | \leq | a_n |$ for all n, then $\| \sum_{n=0}^{\infty} b_n e_n \| \leq$
$K_3 \| \sum_{n=0}^{\infty} a_n e_n \|$.

(e)   For all $x = \sum_{n=0}^{\infty} a_n e_n$ and $\epsilon > 0$, there exists $A_0 \subset \mathbb{N}$, $|A_0| < +\infty$ such that

$$\forall A \supset A_0, \qquad |A| < +\infty, \qquad \left\| x - \sum_{n \in A} a_n e_n \right\| \leq \epsilon$$

Proof.   (b') $\Rightarrow$ (b), (c') $\Rightarrow$ (c), (d') $\Rightarrow$ (d) are obvious; (b) $\Rightarrow$ (b'), (c) $\Rightarrow$ (c'), (d) $\Rightarrow$ (d') are simple applications of the Banach-Steinhaus theorem.

Here, use the family of operators

$$T_{n,\epsilon}\left(\sum_{i=0}^{\infty} a_i e_i \right) = \sum_{i=0}^{n} \epsilon_i a_i e_i \quad \text{for (b)} \Rightarrow \text{(b')}$$

$$T_{n,A}\left(\sum_{i=0}^{\infty} a_i e_i \right) = \sum_{\substack{i=0 \\ i \in A}}^{n} a_i e_i \quad \text{for (c)} \Rightarrow \text{(c')}$$

$$T_{n,(b_n)_{n \in \mathbb{N}}}\left(\sum_{i=0}^{\infty} a_i e_i \right) = \sum_{i=0}^{n} b_i e_i \quad \text{for (d)} \Rightarrow \text{(d')}$$

and conclude as in the proof of Proposition I.1.13.

To prove (b') $\Leftrightarrow$ (c') $\Leftrightarrow$ (d'), fix a finite sequence of real numbers $(a_i)_{i \in \mathbb{N}}$.

(b') $\Rightarrow$ (c'): Let $A \subset \mathbb{N}$ be given. Set $\epsilon_i = +1$ if $i \in A$ and $\epsilon_i = -1$ if $i \notin A$. Then by (b'):

$$\left\| \sum_{i=0}^{\infty} \epsilon_i a_i e_i \right\| \leq K_1 \left\| \sum_{i=0}^{\infty} a_i e_i \right\|$$

So

$$\left\| \sum_{i \in A} a_i e_i - \sum_{i \in A^c} a_i e_i \right\| \leq K_1 \left\| \sum_{i=0}^{\infty} a_i e_i \right\|$$

And thus

$$\left\| \sum_{i \in A} a_i e_i \right\| \leq \left\| \frac{\sum_{i \in A} a_i e_i + \sum_{i \in A^c} a_i e_i}{2} \right\| + \left\| \frac{\sum_{i \in A} a_i e_i - \sum_{i \in A^c} a_i e_i}{2} \right\|$$

$$\leq \frac{K_1 + 1}{2} \left\| \sum_{i=0}^{\infty} a_i e_i \right\|$$

$(c') \Rightarrow (b')$: Let $(\epsilon_i)_{i \in \mathbb{N}}$, $\epsilon_i = +1$ be given and set $A = \{i \in \mathbb{N}, \epsilon_i = +1\}$. Then

$$\left\| \sum_{i=0}^{\infty} \epsilon_i a_i e_i \right\| = \left\| \sum_{i \in A} a_i e_i - \sum_{i \in A^c} a_i e_i \right\| \leq 2K_2 \left\| \sum_{i=0}^{\infty} a_i e_i \right\|$$

$(d') \Rightarrow (b')$ is obvious.

$(b') \Rightarrow (d')$: We need a lemma.

Lemma I.3.3. If $x = \sum_{n=0}^{\infty} a_n e_n$, set $|x| = \text{Sup}\{\| \sum_{n=0}^{\infty} \epsilon_n a_n e_n \|/ \epsilon_n = \pm 1$ for all $n \in \mathbb{N}\}$. Then

1. $|\cdot|$ is equivalent to $\|\cdot\|$ if (b') is satisfied.
2. $(e_n)_{n \in \mathbb{N}}$ verifies (c') with $K_2 = 1$ for $|\cdot|$.
3. If $A, B \subset \mathbb{N}$, $A \cap B = \varnothing$, $|A|, |B| < +\infty$, and if $x = \sum_{n \in A} b_n e_n$, $y = \sum_{n \in B} b_n e_n$, the real function $t \to |x + ty|$ is an even, convex, and continuous function of t.

Proof of Lemma I.3.3

1. This is obvious because if (b') is satisfied, then for all $x \in X$,

$$\|x\| \leq |x| \leq K_1 \|x\|$$

2. If $A \subset \mathbb{N}$ and $x = \sum_{n=0}^{\infty} a_n e_n$, we get

$$|x| \geq \left\| \sum_{n \in A} \epsilon_n a_n e_n + \sum_{n \in A^c} \epsilon_n a_n e_n \right\|$$

and

$$|x| \geq \left\| \sum_{n \in A} \epsilon_n a_n e_n - \sum_{n \in A^c} \epsilon_n a_n e_n \right\|$$

Thus $|x| \geq \| \sum_{n \in A} \epsilon_n a_n e_n \|$ for all $\epsilon_n = \pm 1$ and $|\cdot|$ verifies (c') with $K_2 = 1$.

3. This is obvious.

Let us come back to the proof of (b') $\Rightarrow$ (d'): Set $b_n = t_n a_n$, $|t_n| \leq 1$. Then, by 3. of Lemma I.3.3,

$$\left\| \sum_{n=0}^{\infty} b_n e_n \right\| \leq \left| \sum_{n=0}^{\infty} b_n e_n \right| = \left| \sum_{n=0}^{\infty} a_n t_n e_n \right| \leq \left| \sum_{n=0}^{\infty} a_n \epsilon_n e_n \right|$$

where $\epsilon_n = \text{sgn } t_n$. Then

$$\left\| \sum_{n=0}^{\infty} b_n e_n \right\| \leq \left| \sum_{n=0}^{\infty} a_n \epsilon_n e_n \right| \leq K_1 \left\| \sum_{n=0}^{\infty} a_n e_n \right\|$$

(a) $\Rightarrow$ *(e):* Assume that (e) is false; then there exists $x = \sum_{n=0}^{\infty} a_n e_n$ and $\epsilon > 0$ such that for all $A_0 \subset \mathbb{N}$, $|A_0| < +\infty$, there exists $A \supset A_0$ such that

$$\left\| \sum_{n \in A} a_n e_n - x \right\| \geq \epsilon$$

By induction, build a sequence $(A_n)_{n \in \mathbb{N}}$ in $\mathbb{N}$ such that

$$\begin{cases} A_k \supset (A_{k-1} \cup \{0, 1, \ldots, k\}) \\ \left\| \sum_{n \in A_k} a_n e_n - x \right\| \geq \epsilon \end{cases}$$

Let π be the permutation of $\mathbb{N}$ obtained by enumerating A_0, then $A_1 \backslash A_0$, $A_2 \backslash A_1, \ldots$. The series $\sum_{n=0}^{\infty} a_{\pi(n)} e_{\pi(n)}$ cannot converge to x.

(e) $\Rightarrow$ *(c'):* Let $A \subset \mathbb{N} \mid A \mid < +\infty$, and define $P_A (\sum_{n=0}^{\infty} a_n e_n) = \sum_{n \in A} a_n e_n$. Since $(e_n)_{n \in \mathbb{N}}$ is a basis, P_A is continuous. In order to prove (c') it is sufficient to prove that $\sup\{\| P_A \|, A \subseteq \mathbb{N}, |A| < +\infty\}$ is finite. But by the Banach-Steinhaus theorem, it is enough to prove that, for all fixed x in E, $\sup\{\| P_A x \|, A \subseteq \mathbb{N}, |A| < +\infty\}$ is finite.

If not, for some $x \in X$, we can find $(A_k)_{k \geq 1}$, $A_k \subset \mathbb{N}$, $|A_k| < +\infty$ such that

$$\| P_{A_k} x \| \xrightarrow[k \to +\infty]{} +\infty$$

Apply (e) with $\epsilon = 1$ and this element $x \in X$. Then for the corresponding $A_0 \subset \mathbb{N}$ we get $\| P_{A_k \cup A_0} x \| \leq \| x \| + 1$. But

$$\| P_{A_k \cup A_0} x \| = \| P_{A_k} x + P_{A_0 \backslash A_k} x \| \geq \| P_{A_k} x \| - \| P_{A_0 \backslash A_k} x \|$$

Since $A_0 \backslash A_k \subset A_0$, $\| P_{A_0 \backslash A_k} x \|$ is bounded, and this gives a contradiction.

(c') $\Rightarrow$ *(a):* Let π be a given permutation of $\mathbb{N}$, $\sum_{n=0}^{\infty} a_n e_n$ a convergent series, and $\epsilon > 0$. Choose $P_0 \in \mathbb{N}$ such that $\| \sum_{n=P_0+1}^{\infty} a_n e_n \| \leq \epsilon$. Let N_0 be such that the set $\{\pi(n), n > N_0\}$ does not contain any $n \leq P_0$.

Then, for all $N > N_0$,

$$\left\| \sum_{n=N_0+1}^{N} a_{\pi(n)} e_{\pi(n)} \right\| \leq K_2 \left\| \sum_{n=P_0+1}^{\infty} a_n e_n \right\| \leq K_2 \epsilon \quad \text{by (c')}$$

Thus the series $\sum_{n=0}^{\infty} a_{\pi(n)} e_{\pi(n)}$ converges.

Definition I.3.4

1. The *unconditional basis constant* of $(e_n)_{n \in \mathbb{N}}$ is the smallest K_2 satisfying (c'), that is, such that $\| \sum_{n \in A} a_n e_n \| \leq K_2 \| \sum_{n=0}^{\infty} a_n e_n \|$.
2. If the unconditional basis constant of $(e_n)_{n \in \mathbb{N}}$ is K, we will say that $(e_n)_{n \in \mathbb{N}}$ is *K-unconditional*.

Remark I.3.5

1. When $(e_n)_{n \in \mathbb{N}}$ is unconditional, for all permutations π, the sum $\sum_{n=0}^{\infty} a_{\pi(n)} e_{\pi(n)}$ is the same as $\sum_{n=0}^{\infty} a_n e_n$. On the contrary, the sum $\sum_{n=0}^{\infty} \epsilon_n a_n e_n$ may be different.
2. A block-basis of an unconditional basis is unconditional.

Examples I.3.6

1. The unit vector basis $(e_n)_{n \in \mathbb{N}}$ of ℓ_p $(1 \leq p < +\infty)$ or c_0 is 1-unconditional.
2. The summing basis of c_0 defined by $\forall n \in \mathbb{N}$, $s_n = \sum_{i=0}^{n} e_i$ is not unconditional; it is clear that $\| \sum_{i=0}^{\infty} a_i s_i \|_{c_0} = \text{Sup}_{n \in \mathbb{N}} | \sum_{i=n}^{\infty} a_i |$. Then

$$\begin{cases} \left\| \sum_{i=0}^{n} s_i \right\|_{c_0} = n \\ \left\| \sum_{i=0}^{n} (-1)^i s_i \right\|_{c_0} = 1 \end{cases}$$

and $(s_n)_{n \in \mathbb{N}}$ is not unconditional.

The two following theorems study the existence of unconditional bases in $L_p[0,1]$ $(1 \leq p < \infty)$. The first one, on L_1, is due to A. Pelcynski [Pe1], and we present a proof of it by V. D. Milman [Mi]. See also [LT]. Let us first give a definition.

Definition I.3.7. The sequence of *Rademacher functions* $(r_n)_{n \in \mathbb{N}}$ is defined by

$$r_0(t) = +1 \qquad t \in [0,1[$$

$$r_1(t) = +1 \qquad t \in [0,\tfrac{1}{2}[$$

$$ = -1 \qquad t \in [\tfrac{1}{2},1[$$

$$r_2(t) = +1 \qquad t \in [0,\tfrac{1}{4}[\cup [\tfrac{1}{2},\tfrac{3}{4}[$$

$$ = -1 \qquad t \in [\tfrac{1}{4},\tfrac{1}{2}[\cup [\tfrac{3}{4},1[$$

$$\vdots$$

$$k \in \mathbb{N}, \qquad r_{k+1}(t) = +1 \qquad t \in \bigcup_{\ell=1}^{2^k} \left[\frac{2\ell - 2}{2^{k+1}}, \frac{2\ell - 1}{2^{k+1}} \right[$$

$$\phantom{k \in \mathbb{N}, \qquad r_{k+1}(t)} = -1 \qquad t \in \bigcup_{\ell=1}^{2^k} \left[\frac{2\ell - 1}{2^{k+1}}, \frac{2\ell}{2^{k+1}} \right[$$

$$\vdots$$

Recall that the intervals

$$\left[\frac{2\ell - 1}{2^{k+1}}, \frac{2\ell}{2^{k+1}} \right[$$

for $1 \le k \le 2^n$ and $n \in \mathbb{N}$ are called *dyadic*.

Note that the sequence of Rademacher functions is a sequence of independent and identically distributed random variables on $[0,1]$ such that $\lambda\{r_n = 1\} = \lambda\{r_n = -1\} = \tfrac{1}{2}$, where λ denotes the Lebesgues measure on $[0,1]$.

Theorem I.3.8. $L_1 [0,1]$ is not isomorphic to a subspace of a space with an unconditional basis.

Proof. For $x \in L_1$, we are going to prove that:

(i) $r_n x \xrightarrow[n \to +\infty]{} 0$ for $\sigma(L_1, L_\infty)$

(ii) $\| x + r_n x \| \xrightarrow[n \to +\infty]{} \| x \|$.

Indeed, by the density of the dyadic intervals in the Borel σ-field on $[0,1]$, it is sufficient to prove (i) and (ii) for characteristic functions of such intervals; thus, we can suppose that

$$x = 1\left[\frac{2l-2}{2^{k+1}}, \frac{2l-1}{2^{k+1}}\right[$$

Let us prove (i): If

$$y = 1\left[\frac{2i-2}{2^{j+1}}, \frac{2i-1}{2^{j+1}}\right[$$

then for all n more than k and j, we have

$$\int_0^1 r_n(t)x(t)y(t)\, dt = 0$$

Thus, if g is a linear combination of characteristic functions of dyadic intervals, we get that the integral $\int_0^1 r_n(t)x(t)g(t)\, dt$ tends to 0 when n goes to infinity.

If h belongs to $L_\infty[0,1]$, it also belongs to $L_1[0,1]$ and for all $\epsilon > 0$ there exists $g \in L_1$, a linear combination of characteristic functions of dyadic intervals, such that $\| g - h \|_1 \le \epsilon$. Then we can write

$$\left|\int_0^1 r_nxh\right| \le \int_0^1 |r_nx(h - g)| + \left|\int_0^1 r_nxg\right| \le \epsilon + \left|\int_0^1 r_nxg\right|$$

Since $(\int_0^1 r_nxg)_{n\in\mathbb{N}}$ converges to 0, this proves that $(\int_0^1 r_nxh)_{n\in\mathbb{N}}$ converges to 0 too. This concludes the proof of i).

Let us prove (ii): With the same notation, we can see that if n is more than k, we have

$$\int_0^1 |x + r_nx| = \int_0^1 |1 + r_n|\, x = 2\left\|\frac{x}{2}\right\|_1 = \| x \|_1$$

This proves (ii).

Suppose that $L_1[0,1]$ is a subspace of Y, where Y is a Banach space with an unconditional basis $(e_n)_{n\in\mathbb{N}}$. As in Section 1, let us denote by $(P_n)_{n\in\mathbb{N}}$ the sequence of projections associated with the basis $(e_n)_{n\in\mathbb{N}}$, namely $P_n(\sum_{i=0}^\infty a_ie_i) = (\sum_{i=0}^n a_ie_i)$. Note that each P_n has finite rank and thus is a compact operator.

Let $x_0 \in L_1[0,1]$ be fixed with $\| x_0 \| = 1$. Then there exists $N_0 \in \mathbb{N}$ such that $\| x_0 - P_{N_0}(x_0)\| \le \frac{1}{2}$. With properties (i) and (ii), we are going to construct by induction a sequence $(x_n)_{n\in\mathbb{N}}$ in $L_1[0,1]$ such that:

(iii) For all $n \in \mathbb{N}$, $\frac{1}{2} \le \| x_n \| = \| x_0 + \cdots + x_{n-1} \| \le 2$.

(iv) The sequence $(x_n)_{n \in \mathbb{N}}$ is equivalent to a block-basis of $(e_n)_{n \in \mathbb{N}}$.

By (i), (ii), and the compactness of P_{N_0}, there exists $k_1 \in \mathbb{N}$ such that

$$\| P_{N_0}(r_{k_1} x_0) \| \le 1/2^2 \quad \text{and} \quad \big| \, \| x_0 + r_{k_1} x_0 \| - 1 \, \big| \le 1/2^2$$

We define $x_1 = r_{k_1} x_0$. Then $\| x_1 \| = \| x_0 \| = 1$ and there exists $N_1 \in \mathbb{N}$ such that

$$\| x_1 - P_{N_1}(x_1) \| \le 1/2^2$$

Again by i) and ii) there exists $k_2 \in \mathbb{N}$ such that

$$\| P_{N_1}(r_{k_2}(x_0 + x_1)) \| \le 1/2^3 \quad \text{and}$$
$$\big| \, \| x_0 + x_1 + r_{k_2}(x_0 + x_1) \| - \| x_0 + x_1 \| \, \big| \le 1/2^3$$

We define $x_2 = r_{k_2}(x_0 + x_1)$. Suppose that $x_0, x_1, \ldots, x_{n-1}$ have already been chosen in that way; that is, there exist integers $(k_i)_{1 \le i \le n-1}$ and $(N_i)_{1 \le i \le n-1}$ with the following properties, called (P_i), for all $i = 1, 2, \ldots, n - 1$:

$$\| P_{N_{i-1}}(r_{k_i}(x_0 + \cdots + x_{i-1})) \| \le 1/2^{i+1}$$
$$\big| \, \| x_0 + \cdots + x_{i-1} + r_{k_i}(x_0 + \cdots + x_{i-1}) \|$$
$$\qquad\qquad - \| x_0 + \cdots + x_{i-1} \| \, \big| \le 1/2^{i+1}$$
$$x_i = r_{k_i}(x_0 + \cdots + x_{i-1})$$
$$\| x_i - P_{N_i}(x_i) \| \le 1/2^{i+1}$$

Then by (i) and (ii) there exists $k_n \in \mathbb{N}$ such that

$$\| P_{N_{n-1}}(r_{k_n}(x_0 + \cdots + x_{n-1})) \| \le 1/2^{n+1}$$

and

$$\big| \, \| x_0 + \cdots + x_{n-1} + r_{k_n}(x_0 + \cdots + x_{n-1}) \|$$
$$\qquad\qquad - \| x_0 + \cdots + x_{n-1} \| \, \big| \le 1/2^{n+1}$$

We define $x_n = r_{k_n}(x_0 + \cdots + x_{n-1})$ and there exists $N_n \in \mathbb{N}$ such that

$$\| x_n - P_{N_n}(x_n) \| \le 1/2^{n+1}$$

Thus (P_n) is verified and the induction is proved.

By construction, $(x_n)_{n \in \mathbb{N}}$ has property (iii).

If we define a sequence of blocks $(u_n)_{n \in \mathbb{N}}$ on $(e_n)_{n \in \mathbb{N}}$ by

$$u_n = P_{N_n}(x_n) - P_{N_{n-1}}(x_n)$$

then, for all n we get $\|u_n - x_n\| \leq 1/2^n$, then, by the proof of Proposition I.1.15, and forgetting the first terms if necessary, $(x_n)_{n \in \mathbb{N}}$ and $(u_n)_{n \in \mathbb{N}}$ are equivalent.

So, this sequence $(x_n)_{n \in \mathbb{N}}$ satisfies (iii) and (iv).

Obviously, since $(e_n)_{n \in \mathbb{N}}$ is unconditional by hypothesis, $(x_n)_{n \in \mathbb{N}}$ is also unconditional. Let K be its unconditional basis constant. Then for all finite sequences $(a_i)_{i \in \mathbb{N}}$ we can write

$$\left\| \sum_{i=0}^{\infty} a_i x_i \right\| \leq K \sup_{i \in \mathbb{N}} |a_i| \left\| \sum_{i=0}^{\infty} x_i \right\|$$

$$\leq 2K \sup_{i \in \mathbb{N}} |a_i|$$

and

$$\left\| \sum_{i=0}^{\infty} a_i x_i \right\| \geq \frac{1}{2K} \operatorname{Sup}_{i \in \mathbb{N}} |a_i|$$

Thus $(x_n)_{n \in \mathbb{N}}$ is equivalent to the unit vector basis of c_0.

To prove that this is impossible, it is sufficient to show the following lemma.

Lemma I.3.9.   The space c_0 is not isomorphic to a subspace of L_1.

Proof.   Let T be an isomorphism from c_0 into L_1. If $(e_n)_{n \in \mathbb{N}}$ denotes the unit vector basis of c_0, we note $x_n = Te_n$ for all $n \in \mathbb{N}$.

We recall *Khintchine's inequalities*, which will be proved in Chapter IV (Proposition IV.2.12) but which also figure in [LT] and [Be3].

Khintchine's Inequalities.   Let $(r_n)_{n \in \mathbb{N}}$ be the sequence of Rademacher functions on $[0,1]$, defined above in I.3.7. Then, for all $p \in [1, +\infty]$, there exist positive constants A_p and B_P such that, for all $n \in \mathbb{N}$ and $a_0, \ldots, a_n$ in $\mathbb{R}$:

$$A_P \left(\sum_{i=0}^{n} a_i^2 \right)^{1/2} \leq \left(\int_0^1 \left| \sum_{i=0}^{n} a_i r_i(t) \right|^p dt \right)^{1/P} \leq B_p \left(\sum_{i=0}^{n} a_i^2 \right)^{1/2}$$

Let us apply these inequalities with $p = 1$ and $a_i = x_i(\omega)$ and integrate in $\omega \in [0,1]$:

$$A_1 \int_0^1 \left(\sum_{i=0}^{n} x_i(\omega)^2 \right)^{1/2} d\omega \leq \int_0^1 \left[\int_0^1 \left| \sum_{i=0}^{n} x_i(\omega) r_i(t) \right| dt \right] d\omega$$

$$= \int_0^1 \left\| \sum_{i=0}^n x_i r_i(t) \right\|_1 dt$$

$$\leq \| T \| \left(\int_0^1 \left\| \sum_{i=0}^n e_i r_i(t) \right\|_{co} dt \right)$$

$$= \| T \|$$

On the other hand, we need a lemma to compute $\int_0^1 (\sum_{i=0}^n x_i(\omega)^2)^{1/2} d\omega$ and get a contradiction.

Lemma I.3.10. Let $\alpha \in]0,1[$. If f and g are two positive and measurable functions on $[0,1]$, we have

$$\left(\int_0^1 f^\alpha(\omega) \, d\omega \right)^{1/\alpha} + \left(\int_0^1 g^\alpha(\omega) \, d\omega \right)^{1/\alpha} \leq \left(\int_0^1 (f + g)^\alpha(\omega) \, d\omega \right)^{1/\alpha}$$

Proof of Lemma I.3.10. Let us define $p = 1/\alpha$ and $1/q = 1 - 1/p$. We can write Hölder's inequalities with parameters p and q:

$$\int_0^1 f^\alpha(\omega) \, d\omega = \int_0^1 \frac{f^\alpha}{(f + g)^{\alpha/q}} (f + g)^{\alpha/q}$$

$$\leq \left(\int_0^1 \frac{f}{(f + g)^{1/q}} \right)^{1/p} \left(\int_0^1 (f + g)^\alpha \right)^{1/q}$$

Changing f into g, we get

$$\int_0^1 g^\alpha(\omega) \, d\omega \leq \left(\int_0^1 \frac{g}{(f + g)^{1/q}} \right)^{1/p} \left(\int_0^1 (f + g)^\alpha \right)^{1/q}$$

Taking the power $1/\alpha$ of each side of these inequalities and adding them, we get

$$\left(\int_0^1 f^\alpha(\omega) \, d\omega \right)^{1/\alpha} + \left(\int_0^1 g^\alpha(\omega) \, d\omega \right)^{1/\alpha}$$

$$\leq \int_0^1 (f + g)^{1 - 1/q} \left(\int_0^1 (f + g)^\alpha \right)^{1/q\alpha}$$

$$= \left(\int_0^1 (f + g)^\alpha \right)^{1/\alpha}$$

This proves Lemma I.3.10.

We can apply this inequality with $\alpha = \frac{1}{2}$ and compute $\int_0^1 (\sum_{i=0}^n x_i(\omega)^2)^{1/2} \, d\omega$:

$$\int_0^1 \left(\sum_{i=0}^n x_i(\omega)^2 \right)^{1/2} d\omega \geq \sum_{i=0}^n \left(\int_0^1 |x_i(\omega)| \, d\omega \right)^2$$

$$= \sum_{i=0}^n \|x_i\|_1^2 \geq \frac{n+1}{\|T\|^2}$$

This is a contradiction because we cannot have

$$A_1^2 \frac{n+1}{\|T\|^2} \leq \|T\|^2 \quad \text{for all } n \in \mathbb{N}$$

Thus, $L_1[0,1]$ does not contain any subspace isomorphic to c_0, and this proves Lemma I.3.9 and Theorem I.3.8.

The following result is due to R. Paley [Pa], and here we give a proof of it by D. Burkholder [Bur]. See also [Be3, appendix].

The Haar functions on [0,1] have already appeared in Example I.1.6, where it was shown that this sequence is a monotone basis of $L_p[0,1]$ for $p \in]1, +\infty[$. Here we are going to show that it is an unconditional basis of $L_p[0,1]$ for $p \in]1, +\infty[$. Let us first recall the definition of these functions.

Definition I.3.11. For $k \in \mathbb{N}$ and $1 \leq l \leq 2^k$, we note $n = 2^k + l - 1$ and then the nth Haar function is defined by

$$h_n(\omega) = +1 \quad \text{if } \omega \in \left[\frac{2l-2}{2^{k+1}}, \frac{2l-1}{2^{k+1}} \right]$$

$$= -1 \quad \text{if } \omega \in \left[\frac{2l-1}{2^{k+1}}, \frac{2l}{2^{k+1}} \right]$$

$$= 0 \quad \text{otherwise}$$

Theorem I.3.12. The Haar basis is an unconditional basis of $L_p[0,1]$ $(1 < p < +\infty)$.

More precisely, define

$$p^* = \max\left(p, \frac{p}{p-1} \right)$$

If $(h_n)_{n \in \mathbb{N}}$ are the Haar functions on [0,1], then for all $n \in \mathbb{N}$, $\epsilon_k = \pm 1$ for $k \leq n$, and $(a_k)_{k=0,\dots,n} \in \mathbb{R}^{n+1}$, we have

(*)
$$\left\| \sum_{k=0}^{n} \epsilon_k a_k h_k \right\|_p \leq (p^* - 1) \left\| \sum_{k=0}^{n} a_k h_k \right\|_p$$

Proof. Let $Z_n : [0,1] \to \mathbb{R}^2$ be the function given by $Z_n = (X_n, Y_n)$ where

$$\begin{cases} X_n = \displaystyle\sum_{k=0}^{n} (\epsilon_k + 1) a_k h_k \\ Y_n = \displaystyle\sum_{k=0}^{n} (\epsilon_k - 1) a_k h_k \end{cases}$$

Define $v : \mathbb{R}^2 \to \mathbb{R}$ by

$$v(x,y) = \left| \frac{x+y}{2} \right|^p - (p^* - 1)^p \left| \frac{x-y}{2} \right|^p$$

so that

$$\int_0^1 v(Z_n)(t) \, dt = \left\| \sum_{k=0}^{n} \epsilon_k a_k h_k \right\|_p^p - (p^* - 1)^p \left\| \sum_{k=0}^{n} a_k h_k \right\|_p^p$$

Clearly (*) is equivalent to $\int_0^1 v(Z_n)(t) \, dt \leq 0$.

Suppose for a while that there exists a differentiable function $u : \mathbb{R}^2 \to \mathbb{R}$ such that

(i) For given x and $y \in \mathbb{R}$, $u(.,y)$ and $u(x,.)$ are concave.

(ii) $v \leq u$.

(iii) $u(x,u) \leq 0$ if $xy = 0$.

Then, by the concavity condition (i), we get

$$hk = 0 \Rightarrow u(x + h, y + k) \leq u(x,y) + u_x'(x,y)h + u_y'(x,y)k$$

So, since $(\epsilon_n - 1)(\epsilon_n + 1) = 0$, then $(X_n - X_{n-1})(Y_n - Y_{n-1}) = 0$, and this last inequality applied to (X_n, Y_n) and integrated gives

$$u(Z_n) \leq u(Z_{n-1}) + u_x'(Z_{n-1})(X_n - X_{n-1}) + u_y'(Z_{n-1})(Y_n - Y_{n-1})$$

But both terms $u_x'(Z_{n-1})(X_n - X_{n-1})$ and $u_y'(Z_{n-1})(Y_n - Y_{n-1})$ are null because they are the integrals of the product of h_n by a function of $(h_1, \ldots, h_{n-1})$ and the Haar functions are orthogonal in L_2. So, by induction,

$$u(Z_n) \le u(Z_{n-1}) \le \cdots \le u(Z_0)$$

Since $X_0 Y_0 = 0$, by (iii) $u(Z_0) \le 0$ and so by (ii) we get

$$\forall n \in \mathbb{N}, \qquad v(Z_n) \le 0$$

This proves (*).

Definition of u: In $\mathbb{R}^2$, we divide the first quadrant $\{(x,y) \in \mathbb{R}^2 / x > 0$ and $-x < y < x\}$ into two parts:

$$A_1 = \{(x,y) \in \mathbb{R}^2 / x > 0, (1 - 2/p^*)x < y < x\}$$

$$A_2 = \left\{ (x,y) \in \mathbb{R}^2 / -x < y < \left(1 - \frac{2}{p^*}\right)x \right\}$$

We put $\alpha_p = p[p^*/p^* - 1]^{1-p}$. Let u be the continuous function on $\mathbb{R}^2$ given as follows.
If $1 < p \le 2$:

$$u(x,y) = v(x,y) \qquad\qquad \text{if } (x,y) \in A_1$$

$$u(x,y) = \alpha_p x^p \left[1 - \frac{p^*(x - y)}{2x} \right] \quad \text{if } (x,y) \in A_2$$

If $p > 2$:

$$u(x,y) = \alpha_p x^p \left[1 - \frac{p^*(x - y)}{2x} \right] \quad \text{if } (x,y) \in A_1$$

$$u(x,y) = v(y,x) \qquad\qquad \text{if } (x,y) \in A_2$$

In the other quadrants, we put a symmetry condition:

$$u(x,y) = u(y,x) = u(-x,-y)$$

Lemma I.3.13. The function u defined above has continuous first partial derivatives u'_x and u'_y and satisfies (i), (ii), (iii).

Sketch of Proof. If $p = 2$, this is obvious, because $u(x,y) = xy$.
    Suppose $p > 2$, the case $p < 2$ being similar. The function u is infinitely differentiable away from the four lines:

$$x + y = 0, \qquad x = y, \qquad y = (1 = 2/p^*)x, \qquad x = (1 - 2/p^*)y$$

Fix $x > 0$; then u'_x exists and is continuous at $(x, (1 - 2/p^*)x)$ and

$$u'_x(x, (1 - 2/p^*)x) = \alpha^p x^{p-1}(1 - p^*/2)$$

Similarly, at (x,x), $(x, -x)$, $(x, (1 - 2/p^*)^{-1}x)$, u'_x exists and is continuous. Thus by the symmetry condition, both u'_x and y'_y exist and are continuous on $\mathbb{R}^2$.

In addition, off the four lines above, the second partial derivatives of u satisfy $u''_{xx} \leq 0$, $u''_{yy} \leq 0$, and $u''_{xy} \geq 0$.

Let $hk \leq 0$ with $h + k \neq 0$ and define $g : \mathbb{R} \to \mathbb{R}$ by $g(t) = u(x + ht, y + kt)$. For all but a finite number of values of t, the function g is twice continuously differentiable on a neighborhood of $(x + ht, y + kt)$ and

$$g''(t) = h^2 u''_{xx} + 2hk u''_{xy} + k^2 u''_{yy} \leq 0$$

Therefore, by continuity, g' is nonincreasing on $\mathbb{R}$, so g is concave and (i) follows.

(iii) follows at once from the properties of v and the fact that $u = v$ on the set $\{(x,y)/v(x,y) \leq 0\}$, which includes the closure of the second and fourth quadrants of the plane.

To show (ii), we need to show only that $v(x,y) \leq u(x,y)$ if $(x,y) \in A_1$. Fix $x > 0$ and let $a = (1 - 2/p^*)x$ and $b = x$. Define $f : \mathbb{R} \to \mathbb{R}$ by $f(y) = u(x,y) - v(x,y)$. Then f' is continuous on $[a,b]$ and $f''(y) = -v''_{yy}(x,y)$ for $y \in]a,b[$. It is easy to see that $f(a) = 0$, $f(b) > 0$ since $\alpha_p \geq 1$, $f'(a) = 0$, and there is a number $c \in]a,b[$ such that $f''(y) \geq 0$ if $a < y \leq c$ and $f''(y) \leq 0$ if $c \leq y < b$. Therefore, $f'(y) \geq 0$ if $a \leq y \leq c$, which implies that f is nonnegative on $[a,c]$. Since f is concave on $[c,b]$ and is nonnegative at its end points, f is nonnegative on $[c,b]$. Thus $v \leq u$ on A_1, hence on $\mathbb{R}^2$, and this completes the proof.

Remark. It is known that $(p^* - 1)$ is the smallest constant verifying (*); this is a result of R. Paley [Pa].

4. BANACH SPACES CONTAINING ℓ_1 OR c_0

c_0 and ℓ_p spaces are the simplest examples of spaces with an unconditional basis. In particular, c_0 and ℓ_1 have special properties and so do spaces containing c_0 or ℓ_1. In this section we present these results.

First, we have to define the notion of weakly sequentially complete spaces. These properties are also discussed in [LT] and [Be3].

Definition I.4.1.   A Banach space X is said to be *weakly sequentially complete* (w.s.c.) if every $\sigma(X,X^*)$-Cauchy sequence is $\sigma(X,X^*)$-convergent.

Examples I.4.2

1. If X is reflexive and separable, then X is w.s.c.
2. Any Banach space Y isomorphic to a w.s.c. space X is also w.s.c.
3. c_0 is not w.s.c.; indeed the summing basis $(s_n)_{n \in \mathbb{N}}$ defined by $s_n = \sum_{i=0}^{n} e_i$ is $\sigma(c_0, \ell_1)$-Cauchy and not $\sigma(c_0, \ell_1)$-convergent in c_0.
4. We will see in Section 2 of Chapter IV that L_1 is w.s.c (Corollary IV.2.6).

We can now prove the first result on c_0 and ℓ_1, due to R. C. James [J3]:

Theorem I.4.3.   Let X be a Banach space with an unconditional basis $(e_n)_{n \in \mathbb{N}}$. Then X is reflexive if and only if neither c_0 nor ℓ_1 is isomorphic to a subspace of X.

Proof.   Since c_0 and ℓ_1 are not reflexive, this condition is necessary. In view of Theorem I.2.8., for the converse, it is enough to prove:

Proposition I.4.4.   Let X be a Banach space with an unconditional basis $(e_n)_{n \in \mathbb{N}}$. Then

1. $(e_n)_{n \in \mathbb{N}}$ is shrinking if and only if ℓ_1 is not isomorphic to a subspace of X.
2. $(e_n)_{n \in \mathbb{N}}$ is boundedly complete if and only if c_0 is not isomorphic to a subspace of X or, equivalently, if and only if X is w.s.c.

Proof of Proposition I.4.4.   Let K be the unconditional basis constant of $(e_n)_{n \in \mathbb{N}}$.

Proof of 1: If ℓ_1 embeds in X, then ℓ_∞ is isomorphic to a quotient of X^* and thus X^* cannot be separable. So $(e_n^*)_{n \in \mathbb{N}}$ cannot be a basis of X^*.

Suppose now that $(e_n)_{n\in\mathbb{N}}$ is not shrinking. Then, using Proposition I.2.4, one can show that there exist $\epsilon > 0$, $x^* \in X^*$, and $(u_n)_{n\in\mathbb{N}}$ a normalized block-basis of $(x_n)_{n\in\mathbb{N}}$ such that

$$\| x^* \| = 1 \quad \text{and} \quad x^*(u_i) \geq \epsilon \quad \text{for all } i \in \mathbb{N}$$

Thus:

$$\left\| \sum_{j=0}^{\infty} a_j u_j \right\| \geq x^* \left(\sum_{j=0}^{\infty} a_j u_j \right) \geq \epsilon \left(\sum_{j=0}^{\infty} a_j \right) \quad \text{for all } a_j \geq 0$$

But $(u_n)_{n\in\mathbb{N}}$ being K-unconditional, we get, for all $(a_j)_{j\in\mathbb{N}} \in \mathbb{R}^{(\mathbb{N})}$,

$$\left\| \sum_{j=0}^{\infty} a_j u_j \right\| \geq \frac{\epsilon}{K} \sum_{j=0}^{\infty} | a_j |$$

Obviously,

$$\left\| \sum_{j=0}^{\infty} a_j u_j \right\| \leq \sum_{j=0}^{\infty} | a_j |$$

so $(u_n)_{n\in\mathbb{N}}$ is equivalent to the unit vector basis of ℓ_1.

Proof of 2: Suppose that $(e_n)_{n\in\mathbb{N}}$ is not boundedly complete. Then there exists $(a_n)_{n\in\mathbb{N}} \in \mathbb{R}^{\mathbb{N}}$ such that

$$\begin{cases} \left\| \sum_{i=0}^{n} a_i e_i \right\| \leq 1 \quad \text{for all } n \in \mathbb{N} \\ \left(\sum_{i=0}^{n} a_i e_i \right)_{n\in\mathbb{N}} \quad \text{does not converge} \end{cases}$$

So, by Cauchy's condition, there exist $\epsilon > 0$ and two sequences $(p_j)_{j\in\mathbb{N}}$ and $(q_j)_{j\in\mathbb{N}}$ with $p_1 < q_1 < \cdots < p_j < q_j < \cdots$ such that if we define $u_j = \sum_{i=p_j}^{q_j} a_i e_i$ we get $\| u_j \| \geq \epsilon$ for all $j \in \mathbb{N}$. Thus, for $(\lambda_j)_{j=0,\ldots,m} \in \mathbb{R}^{(m+1)}$:

$$\left\| \sum_{j=0}^{m} \lambda_j u_j \right\| \leq K \operatorname*{Sup}_{j=0,\ldots,m} | \lambda_j | \left\| \sum_{j=0}^{m} u_j \right\| \leq K^2 \operatorname*{Sup}_{j=0,\ldots,m} | \lambda_j |$$

On the other hand,

$$\left\| \sum_{j=0}^{m} \lambda_j u_j \right\| \geq \frac{\epsilon}{K} \operatorname*{Sup}_{j=0,\ldots,m} | \lambda_j |$$

Thus, $(u_n)_{n \in \mathbb{N}}$ is equivalent to the unit vector basis of c_0.

Suppose that $(e_n)_{n \in \mathbb{N}}$ is boundedly complete. To prove that X is w.s.c. we need a lemma:

Lemma I.4.5. Let X be a Banach space with an unconditional basis $(e_n)_{n \in \mathbb{N}}$. Suppose that $(y_i)_{i \in \mathbb{N}}$ is a bounded sequence in X such that $\lim_{i \to +\infty} x^*(y_i)$ exists for all $x^* \in X^*$ and $\lim_{i \to +\infty} e_n^*(y_i) = 0$ for all $n \in \mathbb{N}$.

$$\text{Then } \lim_{i \to +\infty} x^*(y_i) = 0 \text{ for all } x^* \in X^*$$

Proof of Lemma I.4.5. Assume that the conclusion of the lemma is false. Then, taking a subsequence if necessary, there exists $x^* \in X^*$ and $\epsilon > 0$ such that $x^*(y_i) \geq \epsilon$, for all $i \in \mathbb{N}$.

By Proposition I.1.17, since $\lim_{i \to +\infty} e_n^*(y_i) = 0$, there exist a block-basis $(u_n)_{n \in \mathbb{N}}$ of $(e_n)_{n \in \mathbb{N}}$ and a subsequence $(y_{n_k})_{k \in \mathbb{N}}$ of $(y_n)_{n \in \mathbb{N}}$ such that

$$\forall k \in \mathbb{N}, \quad \| y_{n_k} - u_k \| \leq \frac{\epsilon}{2^k}$$

So, if $k \geq 2$, $x^*(u_k) \geq \epsilon/2$. Then, as in the proof of 1, for every finite sequence of positive numbers $(a_k)_{k \in \mathbb{N}}$, we can write

$$\left\| \sum_{k=0}^{\infty} a_k u_k \right\| \geq x^* \left(\sum_{k=0}^{\infty} a_k u_k \right) \geq \frac{\epsilon}{2} \left(\sum_{k=0}^{\infty} a_k \right)$$

Since $(u_k)_{k \in \mathbb{N}}$ is K-unconditional, we get

$$\forall (a_k)_{k \in \mathbb{N}} \text{ in } \mathbb{R}^{(\mathbb{N})}, \quad \left\| \sum_{k=0}^{\infty} a_k u_k \right\| \geq (\epsilon/2K) \sum_{k=0}^{\infty} | a_k |$$

Obviously $\| \sum_{k=0}^{\infty} a_k u_k \| \leq M \sum_{k=0}^{\infty} | a_k |$ for some $M < +\infty$, and $(u_k)_{k \in \mathbb{N}}$ is equivalent to the unit vector basis of ℓ_1. Now, it is clear that $(y_{n_k})_{k \in \mathbb{N}}$ is also equivalent to the unit vector basis of ℓ_1.

Thus, there exists $y^* \in X^*$ such that $y^*(y_{n_k}) = (-1)^k$ for all $k \in \mathbb{N}$, which contradicts the hypothesis, and Lemma I.4.5 is proved.

Coming back to the proof of 2 of Proposition I.4.4, we suppose that $(e_n)_{n \in \mathbb{N}}$ is boundedly complete and we want to show that X is w.s.c. Suppose that $(y_i)_{i \in \mathbb{N}}$ is a sequence in X such that $\lim_{i \to +\infty} x^*(y_i)$ exists for all $x^* \in X^*$. Set $a_n = \lim_{i \to +\infty} e_n^*(y_i)$. Then, for all $m \in \mathbb{N}$,

$$\left\| \sum_{n=0}^{m} a_n e_n \right\| = \lim_{i \to +\infty} \| P_m(y_i) \| \le K \sup_{i \in \mathbb{N}} \| y_i \|$$

where P_m are the canonical projections associated with the basis $(e_n)_{n \in \mathbb{N}}$. Because $(y_i)_{i \in \mathbb{N}}$ is $\sigma(X,X^*)$-convergent and thus norm-bounded, $\sup_{i \in \mathbb{N}} \| y_i \|$ is finite. Thus $\sum_{n=0}^{\infty} a_n e_n$ converges to some $x \in X$ by hypothesis.

Applying Lemma I.4.5 to $(y_i - y)_{i \in \mathbb{N}}$, we get that $(y_i)_{i \in \mathbb{N}}$ converges to y for $\sigma(X,X^*)$. So X is w.s.c. It remains to remark that if X is w.s.c, then it cannot contain any subspace isomorphic to c_0 that is not w.s.c.

This concludes Proposition I.4.4. and Theorem I.4.3.

A consequence of Proposition I.4.4 and the fact ℓ_∞ is not separable is the following:

Corollary I.4.6. Let X be a Banach space with an unconditional basis. Then

1. If X is w.s.c., then X is isomorphic to a dual space.
2. If X^* is separable, it has an unconditional basis.
3. If X^{**} is separable, then X is reflexive.

The next result is due to R. C. James [J4] and solves positively in the case of ℓ_1 and c_0 the so-called *distortion problem* (see the Introduction). That is, if X contains a subspace isomorphic to c_0 or ℓ_1 then it contains an *almost isometric* copy of this space.

Theorem I.4.7. Let X be a Banach space and $\epsilon > 0$.

If ℓ_1 (or c_0) is isomorphic to a subspace of X then for all $\epsilon > 0$, ℓ_1 (or c_0) is $(1 + \epsilon)$-isomorphic to a subspace of X.

Proof. Let $Y \subset X$ be a subspace of X that is isomorphic to ℓ_1 and let T be an isomorphism between Y and ℓ_1. Suppose that $T : Y \to \ell_1$ satisfies

$$\forall x \in Y, \quad \alpha \| x \| \le \| Tx \|_1 \le \| x \|.$$

Fix $\epsilon > 0$ and let $(P_n)_{n \in \mathbb{N}}$ be the canonical projections associated with the unit vector basis $(e_n)_{n \in \mathbb{N}}$ of ℓ_1. For all $n \in \mathbb{N}$, define

$$\lambda_n = \sup\{ \| Tx \|_1 / \| x \| + 1, \ P_n(Tx) = 0$$
$$\text{and } Tx \text{ is finitely supported on } (e_n)_{n \in \mathbb{N}} \}$$

Then $(\lambda_n)_{n\in\mathbb{N}}$ decreases to some λ and $\alpha \leq \lambda \leq 1$. Let $n_0 \in \mathbb{N}$ be such that

$$\lambda \leq \lambda_{n_0} \leq \lambda(1 + \epsilon)$$

By definition of $(\lambda_n)_{n\in\mathbb{N}}$, it is clear that there exists a block-basis $(u_n)_{n\in\mathbb{N}}$ of $(e_n)_{n\in\mathbb{N}}$ such that for all $n \in \mathbb{N}$, u_n is the image by T of some norm one vector x_n, $P_{n_0}(u_n) = 0$, and $\| u_n \| \geq \lambda/(1 + \epsilon)$. Then, for all finite sequence of real numbers $(a_k)_{k\in\mathbb{N}}$, we get

$$P_{n_0} \left(\sum_{k=0}^{\infty} a_k u_k \right) = 0$$

Thus

$$\left\| \sum_{k=0}^{\infty} a_k x_k \right\| \geq \frac{1}{\lambda_{n_0}} \left\| \sum_{k=0}^{\infty} a_k u_k \right\|_1 \geq \frac{1}{\lambda_{n_0}} \sum_{k=0}^{\infty} | a_k | \, \| u_k \|$$

(since $(u_k)_{k\in\mathbb{N}}$ is a block-basis of $(e_n)_{n\in\mathbb{N}}$)

$$\geq \frac{1}{\lambda_{n_0}} \frac{\lambda}{1 + \epsilon} \sum_{k=0}^{\infty} | a_k | \geq \frac{1}{(1 + \epsilon)^2} \sum_{k=0}^{\infty} | a_k |$$

Obviously

$$\left\| \sum_{k=0}^{\infty} a_k x_k \right\| \leq \sum_{k=0}^{\infty} | a_k | \, \| x_k \| = \sum_{k=0}^{\infty} | a_k |$$

Thus $\overline{\mathrm{Span}}[x_k, \, k \in \mathbb{N}]$ is $(1 + \epsilon)^2$-isomorphic to ℓ_1, and this gives the result if one chooses ϵ.

We prove the result in the case of c_0 in the same way, setting, if S is an isomorphism from c_0 onto a subspace Y of X,

$$\mu_n = \{\| Sx \| / \| x \|_0 = 1, P_n x = 0 \text{ and } x \text{ is finitely supported on } (e_n)_{n\in\mathbb{N}}\}$$

We can choose n_0 such that $\mu \leq \mu_{n_0} \leq \mu(1 + \epsilon)$. Indeed, there exists a block-basis $(x_n)_{n\in\mathbb{N}}$ on $(e_n)_{n\in\mathbb{N}}$ such that for all $n \in \mathbb{N}$:

$$\| x_n \| = 1, \qquad P_{n_0}(x_n) = 0, \qquad \| Sx \| \geq \frac{\mu}{1 + \epsilon}$$

Then we can write, for all finite sequences of scalars $(a_k)_{k\in\mathbb{N}}$,

$$\left\| \sum_{k=0}^{\infty} a_k Sx_k \right\| \leq \mu_0 \left\| \sum_{k=0}^{\infty} a_k x_k \right\| = \mu_0 \sup_{k\in\mathbb{N}} | a_k |$$

$$\text{(since } (x_k)_{k\in\mathbb{N}} \text{ is a block-basis of } (e_n)_{n\in\mathbb{N}})$$

$$\le \mu(1 + \epsilon) \underset{k\in\mathbb{N}}{\text{Sup}} \, | \, a_k \, |$$

On the other hand, if k_0 is such that $| \, a_{k_0} \, | = \text{Sup}_{k\in\mathbb{N}} | \, a_k \, |$,

$$\frac{2\mu}{1 + \epsilon} \underset{k\in\mathbb{N}}{\text{Sup}} \, | \, a_k \, | \le \| \, 2a_{k_0} S x_{k_0} \, \|$$

$$= \left\| \sum_{k=0}^{\infty} a_k S x_k + a_{k_0} S x_{k_0} - \sum_{\substack{k=0 \\ k\ne k_0}}^{\infty} a_k S x_k \right\|$$

$$\le \left\| \sum_{k=0}^{\infty} a_k S x_k \right\| + \mu(1 + \epsilon) \underset{k\in\mathbb{N}}{\text{Sup}} \, | \, a_k \, |$$

Thus

$$\left\| \sum_{k=0}^{\infty} a_k S x_k \right\| \ge \mu \left(\frac{2}{(1 + \epsilon)} - (1 + \epsilon) \right) \underset{k\in\mathbb{N}}{\text{Sup}} \, | \, a_k \, |$$

This proves our assumption by scaling and adapting ϵ.

Remark. The solution to the general distorsion problem is negative. Namely, there exists a renorming of ℓ_2 which does not contain an almost isometric copy of ℓ_2., This is a very recent result of E. Odell and T. Schlumprecht.

The following theorem on spaces containing c_0 is due to C. Bessaga and A. Pelczynski [BP].

Theorem I.4.8. Let X be a Banach space. The following two properties are equivalent:
   (i) c_0 is isomorphic to a subspace of X.
   (ii) There exists $(x_n)_{n\in\mathbb{N}} \subset X$ such that the following two conditions hold:

$$\begin{cases} \forall x^* \in X^*, \quad \sum_{n=0}^{\infty} | \, x^*(x_n) \, | < +\infty \\ \left(\sum_{n=0}^{N} x_n \right)_{N\in\mathbb{N}} \text{ does not converge in } X. \end{cases}$$

Proof. If (i) is realized and if $(e_n)_{n\in\mathbb{N}}$ denotes the unit vector basis of c_0, then it is obvious that $(e_n)_{n\in\mathbb{N}}$ verifies (ii) in c_0. So its image in X also verifies (ii) and (ii) is true.

Suppose now that (ii) is realized and let $(x_n)_{n \in \mathbb{N}}$ in X be given by (ii). Set $T_{N,\epsilon}(x^*) = \sum_{n=0}^{N} \epsilon_n x^*(x_n)$ where $N \in \mathbb{N}$ and $\epsilon_n = \pm 1$ for all $n \in \mathbb{N}$. Then, for each N and $\epsilon = (\epsilon_n)_{n \in \mathbb{N}}$, $T_{N,\epsilon}$ is a bounded linear operator from X^* to $\mathbb{R}$ and by hypothesis, for fixed $x^* \in X^*$, we have

$$\operatorname*{Sup}_{N,\epsilon} | T_{N,\epsilon} x^* | < +\infty$$

Thus by the Banach-Steinhaus theorem, there is an $M > 0$ such that

$$\left\{ \operatorname*{Sup}_{\| x^* \| \leq 1} \operatorname*{Sup}_{N,\epsilon} \left| \sum_{n=0}^{N} \epsilon_n x^*(x_n) \right| \right\} \leq M < +\infty$$

This implies that

$$\forall x^* \in X^*, \qquad \sum_{n=0}^{\infty} | x^*(x_n)| \leq M \| x^* \|$$

On the other hand, the series $\sum_{n=0}^{\infty} x_n$ diverges. So there exist $\epsilon > 0$ and two sequences $(p_k)_{k \in \mathbb{N}}$ and $(q_k)_{k \in \mathbb{N}}$ with $p_1 < q_1 < \cdots < p_k < q_k < \cdots$ such that

$$\left\| \sum_{i=p_k}^{q_k} x_i \right\| \geq \epsilon \quad \text{for all } k \in \mathbb{N}$$

Set $y_k = \sum_{i=p_k}^{q_k} x_i$. Since $\sum_{k=0}^{\infty} | x^*(y_k)| < +\infty$ for all $x^* \in X^*$ by the first part of the proof, $(y_k)_{k \in \mathbb{N}}$ is $\sigma(X, X^*)$-convergent to 0. Passing to a subsequence, we can suppose that $(y_k)_{k \in \mathbb{N}}$ is a basic sequence (see Theorem I.1.10). Let K be its basic constant. Then, for all finite sequences of real numbers $(a_k)_{k \in \mathbb{N}}$, we can write

$$\left\| \sum_{k=0}^{\infty} a_k y_k \right\| \geq \frac{\epsilon}{2K} \operatorname*{Sup}_{k \in \mathbb{N}} | a_k |$$

and

$$\left\| \sum_{k=0}^{\infty} a_k y_k \right\| \leq \operatorname*{Sup} \left\{ \sum_{k=0}^{\infty} a_k x^*(y_k) / \| x^* \| \leq 1 \right\} \leq M \operatorname*{Sup}_{k \in \mathbb{N}} | a_k |$$

Then $(y_k)_{k \in \mathbb{N}}$ is equivalent to the unit vector basis of c_0 and this proves (i). $\blacksquare$

We now pass to a remarkable extension property of c_0 in separable Banach spaces, namely the following theorem, due to A. Sobczyk [So].

The proof that we give here is due to W. Veech [V] and can be found in [LT], vol. I, p. 106.

Theorem I.4.9. Let c_0 be a subspace of a Banach space X. Then there exists a projection P from X to c_0, of norm at most 2.

Proof. Let $(e_n^*)_{n \in \mathbb{N}}$ be the unit vector basis of $\ell_1 = c_0^*$. Every element e_n^* is a linear functional on c_0, and since c_0 is by hypothesis a subspace of X, e_n^* can be extended to X by the Hahn-Banach theorem. We then get a linear functional x_n^* on X of norm 1.

Define $E = \{x^* \in X^* / \| x^* \| = 1$ and for all x in c_0, $x^*(x) = 0\}$. Since X is separable, B_x^* is metrizable for the $\sigma(X^*, X)$-topology [Br]. Call d a distance on B_X^* that induces this topology. The family $(d(x_n^*, E))_{n \in \mathbb{N}}$ is then a bounded sequence of real numbers. The only limit point of this sequence is 0. Indeed, if x belongs to c_0, then $\sigma(X^*, X) - \lim_n x_n^*(x) = \sigma(X^*, X) - \lim_n e_n^*(x) = 0$. Thus the sequence $(d(x_n^*, E))_{n \in \mathbb{N}}$ converges to 0. This means that the sequence $(x_n - y_n)_{n \in \mathbb{N}}$ is $\sigma(X^*, X)$-convergent to 0.

We can now define the projection P: If x belongs to X, define $Px = (x_1^*(x) - y_1^*(x), x_2^*(x) - y_2^*(x), \ldots)$. Then Px belongs to c_0 and $\| Px \| = \operatorname{Sup}_n \| x_n^*(x) - y_n^*(x) \| \leq 2\| x \|$. Moreover, if x belongs to c_0, we have $x_n^*(x) - y_n^*(x) = x_n^*(x)$. So $Px = x$ and this projection is suitable.

The next result, due to H. P. Rosenthal [Ros3], is fundamental in this theory. However, it is also among the most difficult ones. The proof that we give here requires a result from combinatorics, namely, an infinite extension of Ramsey's theorem. We chose to present it for the sake of completeness. Another proof of this theorem can be found in the book of J. Diestel [Di2].

It is in [F] that J. Faharat remarked that one could use Ramsey's theorem to simplify the proof of Rosenthal's theorem.

Theorem I.4.10. Let $(x_n)_{n \in \mathbb{N}}$ be a bounded sequence in a Banach space X. Then $(x_n)_{n \in \mathbb{N}}$ has a subsequence $(x_{n_k})_{k \in \mathbb{N}}$ satisfying one of the two mutually exclusive conditions:

(i) $(x_{n_k})_{k \in \mathbb{N}}$ is equivalent to the unit vector basis of ℓ_1.
(ii) $(x_{n_k})_{k \in \mathbb{N}}$ is $\sigma(X, X^* \searrow$-Cauchy.

To prove this theorem, we need an infinite extension of Ramsey's theorem which is due to Nash–Williams [N–W]. The initial version of it was first proved in [R]. See also [GP] for another proof and further extensions. The proof that we present here was shown to us by A. Louveau.

Recall first that $\mathcal{P}(\mathbb{N})$ is naturally identified with $\{0,1\}^{\mathbb{N}}$: each sequence of integers $(n_k)_{k \in \mathbb{N}}$ is identified with the sequence of 0 and 1, which is null except at the n_k-ranks.

Theorem I.4.11 (Ramsey). Let $\mathcal{P}_\infty(\mathbb{N})$ be the set of all infinite subsets of $\mathbb{N}$ equipped with the product topology on $\{0,1\}^{\mathbb{N}}$. Suppose that $\mathcal{P}_\infty(\mathbb{N})$ has been written as $\mathcal{O} \cup \mathcal{F}$ where $\mathcal{O}$ is open, $\mathcal{F}$ is closed, and $\mathcal{F} \cap \mathcal{O} = \varnothing$. Then there exists $I \subset \mathbb{N}$, $|I| = +\infty$, such that $\mathcal{P}_\infty(I)$ is included in $\mathcal{O}$ or in $\mathcal{F}$.

Proof. All subsets of $\mathbb{N}$ are going to be considered as ordered sets with the natural order induced by $\mathbb{N}$. We will need to fix some notation, which corresponds to the natural set operations on $\mathcal{P}(\mathbb{N})$, but with the additional structure of order.

Notation. If $u \in \mathcal{P}(\mathbb{N})$, recall that Min u (resp. Max u) denotes the smallest (resp. largest) integer belonging to u and $|u|$ denotes the numbers of distinct elements of u.

If $u \in \mathcal{P}(\mathbb{N})$, is finite, for $v \in \mathcal{P}(\mathbb{N})$, we will note $u \leq v$ if

$$v \cap \{0, 1, \ldots, \text{Max } u\} = u$$

Moreover, if Max $u <$ Min v, $u^\frown v$ will denote the ordered set $u \cup v$.

Let $\mathcal{U}$ be an ultrafilter on $\mathbb{N}$.

Definition I.4.12. If $s \in \mathcal{P}(\mathbb{N})$ is finite, a subset T of $\mathcal{P}_\infty(\mathbb{N})$ is said to be a $\mathcal{U}$-*tree of trunk s* if:

1. $T \neq \varnothing$.
2. $\forall u \in T, s \leq u$.
3. Suppose that a finite set u in $\mathbb{N}$ is such that there exists an infinite set v in $\mathbb{N}$ with Max $u <$ Min v and $u^\frown v \in T$. Then the set of integers n such that Max $u < n$ and $n =$ Min w with $u^\frown w \in T$ belongs to $\mathcal{U}$.

This last property means that when we follow a "branch" of T (that is, an infinite ordered subset of $\mathbb{N}$ belonging to T), at each step the set of integers that can be chosen to stay on a branch of T at the next step belongs to the ultrafilter $\mathcal{U}$.

Then let $\mathcal{M}$ be defined by:

$$\mathcal{M} = \{s \in \mathcal{P}(\mathbb{N}), \mid s \mid < +\infty / \exists T \, \mathcal{U}\text{-tree of trunk } s \text{ such that } T \subset \mathcal{O}\}$$

Lemma I.4.13. Let s be a finite ordered set in $\mathbb{N}$ and define

$$U = \{n \in \mathbb{N} / n > \text{Max } s \text{ and } s^\frown\{n\} \in \mathcal{M}\}$$

If U belongs to $\mathcal{U}$, then s belongs to $\mathcal{M}$.

Proof of Lemma I.4.13. Suppose that $U \in \mathcal{U}$.

For all $k \in U$, let T_k be a $\mathcal{U}$-tree of trunk $s^\frown\{k\}$, belonging to $\mathcal{O}$. Then $s^\frown\cup_{k \in U} T_k$ is a $\mathcal{U}$-tree of trunk s belonging to $\mathcal{O}$. Thus $s \in \mathcal{M}$.

We come back to the proof of Theorem I.4.11. Let us distinguish two cases, depending on the set $\mathcal{M}$:

First case: $\varnothing \in \mathcal{M}$; then there exists a $\mathcal{U}$-tree T of trunk $\varnothing$ belonging to $\mathcal{O}$. That means that there exists $U_\varnothing \in \mathcal{U}$ such that for all $n \in U_\varnothing$, there exists $v \in \mathcal{P}(\mathbb{N})$ with Min $v > n$ and such that $\{n\}^\frown v \in T$.

Fix $n_1 \in U_\varnothing$. Then by the definition of a $\mathcal{U}$-tree (I.4.12), there exists $U_{\{1\}} \in \mathcal{U}$ such that for all $n \in U_{\{1\}}$, there exists $v \in \mathcal{P}(\mathbb{N})$ so that

$$\text{Min } v > n \quad \text{and} \quad \{n_1\}^\frown\{n\}^\frown v \in T \subset \mathcal{O}$$

Fix $n_2 \in U_\varnothing \cap U_{\{1\}}$. Then there exist $U_{\{1,2\}}$ and $U_{\{2\}} \in \mathcal{U}$ such that for all $n \in U_{\{1,2\}}$ and $n' \in U_{\{2\}}$, there exist v and $v' \in \mathcal{P}_\infty(\mathbb{N})$ such that

$$\text{Min } v > n \quad \text{and} \quad \min v' > n'$$

$$\{n_1\}^\frown\{n_2\}^\frown\{n\}^\frown v \in T \quad \text{and} \quad \{n_2\}^\frown\{n'\}^\frown v' \in T$$

Fix $n_3 \in U_\varnothing \cap U_{\{1\}} \cap U_{\{1,2\}} \cap U_{\{2\}}$. Suppose that $n_1, \ldots, n_k$ are already built that way. Then we can take n_{k+1} in $\cap_{s \subset \{1,\ldots,n_k\}} U_s$, where

$$U_s = \{n \in \mathbb{N} / n > \text{Max } s \text{ and}$$

$$\exists v \in \mathcal{P}_\infty(\mathbb{N}) \text{ with Min } v > n \text{ such that } s^\frown\{n\}^\frown v \in T\}$$

Then, by construction $\mathcal{P}_\infty(\{n_1, n_2, \ldots, n_k, \ldots\})$ is included in $\mathcal{O}$. Thus the set $I = \{n_1, n_2, \ldots, n_k, \ldots\}$ is suitable.

Second case: $\varnothing \notin \mathcal{M}$; then the set $\{n \in \mathbb{N}/\{n\} \in \mathcal{M}\}$ does not belong to $\mathcal{U}$ by Lemma I.4.13 and thus the complementary set $U'_\varnothing = \{n \in \mathbb{N}/\{n\} \notin \mathcal{M}\}$ belongs to $\mathcal{U}$.

If $\{n\} \notin \mathcal{M}$, then the set $U'_{\{n\}} = \{m \in \mathbb{N}/\{n\}^\wedge\{m\} \notin \mathcal{M}\}$ belongs to $\mathcal{U}$ by Lemma I.4.13.

More generally, if $s \notin \mathcal{M}$, then the set $U'_s = \{m \in \mathbb{N}/s^\wedge\{m\} \notin \mathcal{M}\}$ belongs to $\mathcal{U}$. Define T by

$u \in T$ if for all decompositions $u = s^\wedge\{n\}^\wedge v$, with s finite and Max s $< n <$ Min v, we have $n \in U'_s$.

Then T is a $\mathcal{U}$-tree of trunk $\varnothing$.

Take $\{n_1, n_2, \ldots, n_k, \ldots\}$ as in the first case; that is,

$$n_{k+1} \in \bigcap_{s \subset \{n_1, \ldots, n_k\}} U'_s$$

and set $I = \{n_1, \ldots, n_k, \ldots\}$ Then by construction, for any finite set s in I, there exists $v \in \mathcal{P}_\infty(\mathbb{N})$, with Min $v >$ Max s, such that $s^\wedge v \in \mathcal{F}$. Since $\mathcal{F}$ is closed, this proves that any infinite subsequence of I belongs to $\mathcal{F}$. Thus $\mathcal{P}_\infty(I)$ is included in $\mathcal{F}$ and I is suitable. This concludes Ramsey's theorem.

Before starting the proof of Theorem I.4.10, we need a definition and a lemma that explains how it is possible to recognize a sequence equivalent to the unit vector basis of ℓ_1 in $C(\Omega$ (see also [Di2]).

Definition I.4.14.   Let $(E_n, O_n)_{n \in \mathbb{N}}$ be a sequence of couples of pairwise disjoint subsets of a compact set Ω. If, for all N and P disjoint subsets of $\mathbb{N}$, we have

$$\bigcap_{k \in N} E_k \cap \bigcap_{k \in P} O_k \neq \varnothing$$

we say that $(E_n, O_n)_{n \in \mathbb{N}}$ is *independent*.

Lemma I.4.15 [Ro3].   Let $(E_n, O_n)_{n \in \mathbb{N}}$ be a sequence of independent subsets of a fixed compact set Ω. Suppose that there exist $r \in \mathbb{R}$, $\delta > 0$, and a bounded sequence $(b_n)_{n \in \mathbb{N}}$ in $C(\Omega)$ such that

$$\left[\begin{array}{l} \forall \omega \in E_n, \ b_n(\omega) \geq r + \delta \\ \forall \omega \in O_n, \ b_n(\omega) \leq r \end{array} \right.$$

Then $(b_n)_{n\in\mathbb{N}}$ is equivalent to the unit vector basis of ℓ_1 in $C(\Omega)$.

Proof of Lemma I.4.15. Fix $a_0 \cdots a_n \in \mathbb{R}^{n+1}$. Obviously,

$$\left\| \sum_{i=0}^{n} a_i b_i \right\| \leq \operatorname*{Sup}_{i\in\mathbb{N}} \| b_i \| \sum_{i=0}^{n} | a_i |$$

For the other inequality, set

$$P = \{i \in \{0, \ldots, n\} / a_i > 0\}$$

$$N = \{i \in \{0, \ldots, n\} / a_i < 0\}$$

We have to consider four cases, depending on the values of r and δ:

$r < 0 < r + \delta$: By hypothesis, there exists $\omega \in \cap_{k\in P} E_k \cap \cap_{k\in N} O_k$. Then

$$\left\| \sum_{k=0}^{n} a_k b_k \right\| \geq \left| \sum_{k=0}^{n} a_k b_k(\omega) \right| \geq (r + \delta) \sum_{k\in P} a_k + | r | \sum_{k\in N} | a_k |$$

$$\geq \operatorname{Inf}\{(r + \delta), | r |\} \sum_{k=0}^{n} | a_k |$$

Thus $(b_n)_{n\in\mathbb{N}}$ is equivalent to the unit vector basis of ℓ_1.

$0 \leq r < r + \delta$: By hypothesis, there exists

$$\begin{cases} \omega_1 \in \bigcap_{k\in P} E_k \cap \bigcap_{k\in N} O_k \\ \omega_2 \in \bigcap_{k\in N} E_k \cap \bigcap_{k\in P} O_k \end{cases}$$

Thus:

$$\sum_{k=0}^{n} a_k b_k(\omega_1) = \sum_{k\in P} a_k b_k(\omega_1) + \sum_{k\in N} a_k b_k(\omega_1)$$

$$\geq (r + \delta) \sum_{k\in P} a_k + \sum_{\substack{k\in N \\ b_k(\omega_1)>0}} a_k b_k(\omega_1)$$

$$\geq (r + \delta) \sum_{k\in P} a_k + r \sum_{\substack{k\in N \\ b_k(\omega_1)>0}} a_k$$

$$\geq (r + \delta) \sum_{k\in P} | a_k | - r \sum_{k\in N} | a_k |$$

In the same manner, we also get

$$\sum_{k=0}^{n} a_k b_k(\omega_2) \geq (r + \delta) \sum_{k\in N} |a_k| - r \sum_{k\in P} |a_k|$$

Summing these two inequalities, we obtain

$$\sum_{k=0}^{n} a_k b_k(\omega_1) + \sum_{k=0}^{n} a_k b_k(\omega_2) \geq \delta \sum_{k=0}^{n} |a_k|$$

So one of the terms on the left-hand side is more than $(\delta/2)$ $\sum_{k=0}^{n} |a_k|$ and this proves that $\|\sum_{k=0}^{n} a_k b_k\| \geq (\delta/2) \sum_{k=0}^{n} |a_k|$. This implies that $(b_n)_{n\in\mathbb{N}}$ is equivalent to the unit vector basis of ℓ_1.

$r < r + \delta < 0$: We change $(b_n)_{n\in\mathbb{N}}$ to $(-b_n)_{n\in\mathbb{N}}$ and we are in the same situation as in the preceding case.

$r + \delta = 0$: We change δ to $\delta/2$ and we are again in the same situation as in the preceding case.

This concludes Lemma I.4.15.

Proof of Theorem I.4.10.   Suppose that $(x_n)_{n\in\mathbb{N}}$ is a bounded sequence in a Banach space X, with no $\sigma(X,X^*)$-Cauchy subsequence. We want to extract a subsequence of $(x_n)_{n\in\mathbb{N}}$ that is equivalent to the unit vector basis of ℓ_1. For this, we are going to consider X as a subset of $C(\overline{B}_{X^*}, \sigma(X^*,X))$. Note that $\overline{B}_{X^*}$ is compact for the topology $\sigma(X^*,X)$ and that the hypothesis on $(x_n)_{n\in\mathbb{N}}$ means that $(x_n)_{n\in\mathbb{N}}$ has no pointwise Cauchy subsequence in $C(\overline{B}_{x^*})$.

We are going to use Lemma I.4.15. We need to divide the proof into two steps:

First step: Let $\mathcal{I} = (I_k^1, I_k^2)_{k\in\mathbb{N}}$ be the set of all couples of open intervals in $\mathbb{R}$ with rational center and radius and such that

$$\text{diam } I_k^1 = \text{diam } I_k^2 \leq \frac{1}{2} \text{ dist}(I_k^1, I_k^2)$$

Then we are going to prove that there exist $k \in \mathbb{N}$ and $P \subset \mathbb{N}$, $|P| = +\infty$, such that for all $M \subset P$, $|M| = +\infty$, there exists $x_M^* \in B_{X^*}$ such that the real sequence $[x_M^*(x_m)]_{m\in M}$ has cluster points in I_k^1 and I_k^2.

Indeed, if not, for all $j \in \mathbb{N}$ and $P \subset \mathbb{N}$, $|P| = +\infty$, there exists M

$\subset P$, $|M| = +\infty$ such that for all $x^* \in B_{X^*}$, $(x^*(x_m))_{m \in M}$ has no cluster point in I_j^1 or in I_j^2.

By induction, taking $j = 0, 1, 2, \ldots$, we get a decreasing sequence $(M_j)_{j \in \mathbb{N}}$ of infinite subsets of $\mathbb{N}$ such that for all $j \in \mathbb{N}$ and $x^* \in B_{X^*}$, $(x^*(x_m))_{m \in M_j}$ has no cluster point in I_j^1 or I_j^2.

Let $(p_j)_{j \in \mathbb{N}}$ be an increasing sequence of $\mathbb{N}$ such that for all j, $p_j \in M_j$. Since $(x_{p_j})_{j \in \mathbb{N}}$ is not pointwise convergent, by assumption on $(x_n)_{n \in \mathbb{N}}$ there exist $x_0^* \in B_{X^*}$ and $d^1 \neq d^2 \in \mathbb{R}$ such that $(x_0^*(x_{p_j}))_{j \in \mathbb{N}}$ has d^1 and d^2 as cluster points. But there exists $j_0 > 0$ such that $d^1 \in I_{j_0}^1$ and $d^2 \in I_{j_0}^2$. Thus the sequence $(x_0^*(x_{p_j}))_{j \geq j_0}$ has cluster points in $I_{j_0}^1$ and $I_{j_0}^2$ and this contradicts the definition of $(p_j)_{j \in \mathbb{N}}$. So the claim of the first step is proved.

In the sequel, call $I_k^1 = I^1$, $I_k^2 = I^2$. By extracting a subsequence and changing notation, we can suppose that $P = \mathbb{N}$.

Second step: Define, for all $n \in \mathbb{N}$,

$$E_n = \{x^* \in \overline{B}_{X^*} / x^*(x_n) \in I^1\}$$

$$-E_n = \{x^* \in \overline{B}_{X^*} / x^*(x_n) \in I^2\}$$

Then we are going to prove that there exists a subsequence $(-E_{k_\ell}, E_{k_\ell})_{\ell \in \mathbb{N}}$ that is independent in $\overline{B}_{X^*}$.

Let $\mathcal{P}_k$ be the collection of sequences $(n_\ell)_{\ell \in \mathbb{N}}$ in $\mathbb{N}$ such that

$$\bigcap_{\ell=0}^{k} (-1)^\ell E_{n_\ell} \neq \varnothing$$

As in Theorem I.4.11, we consider $\mathcal{P}(\mathbb{N})$ as $\{0,1\}^{\mathbb{N}}$. Then we are going to prove that $\bigcap_{k \in \mathbb{N}} \mathcal{P}_k$ is closed in the product topology of $\{0,1\}^{\mathbb{N}}$. Indeed it suffices to prove that each $\mathcal{P}_k$ is closed. Let

$$\begin{cases} (n_\ell^0)_{\ell \in \mathbb{N}} \in \overline{\mathcal{P}}_k \\ B = \{(\epsilon_j)_{j \in \mathbb{N}} \in \{0,1\}^{\mathbb{N}} / \epsilon_j = 1 \text{ if } j \in \{n_1^0, \ldots n_k^0\}\} \end{cases}$$

Then B is a neighborhood of $(n_\ell^0)_{\ell \in \mathbb{N}}$ in $\{0,1\}^{\mathbb{N}}$; thus there exists $(n_\ell^1)_{\ell \in \mathbb{N}}$ in $B \cap \mathcal{P}_k$, that is, which satisfies $n_\ell^1 = n_\ell^0$ for $\ell \leq k$. So

$$\bigcap_{\ell=0}^{k} (-1)^\ell E_{n_\ell^0} = \bigcap_{\ell=0}^{k} (-1)^\ell E_{n_\ell^1} \neq \varnothing \quad \text{and} \quad (n_\ell^0)_{\ell \in \mathbb{N}} \in \mathcal{P}_k$$

This implies that $\mathcal{P}_k = \overline{\mathcal{P}}_k$ and proves our assertion.

We are in position to apply Theorem I.4.11, with $\mathscr{F} = \cap_{k=0}^{\infty} \mathscr{P}_k$ and $\mathbb{O} = (\cap_{k=0}^{\infty} \mathscr{P}_k)^c$. There exists an increasing sequence $(m_\ell)_{\ell \in \mathbb{N}}$ in $\mathbb{N}$ such that

$$\mathscr{P}_\infty((m_\ell)_{\ell \in \mathbb{N}}) \subset \bigcap_{k=0}^{\infty} \mathscr{P}_k \quad \text{or} \quad \left(\bigcap_{k=0}^{\infty} \mathscr{P}_k\right)^c$$

But if we apply the first step of that proof to $M = (m_\ell)_{\ell \in \mathbb{N}}$, we get the existence of $x^* \in B_{X^*}$ such that $x^*(x_{m_\ell})_{\ell \in \mathbb{N}}$ has cluster points in I^1 and in I^2.

Since, I^1 and I^2 are open, we can extract a further subsequence $(x_{m_{\ell_q}})_{q \in \mathbb{N}}$ such that

$$q \text{ even} \Rightarrow x^*(x_{m_{\ell_q}})_{q \in \mathbb{N}} \in I^1$$

$$q \text{ odd} \Rightarrow x^*(x_{m_{\ell_q}})_{q \in \mathbb{N}} \in I^2$$

Thus $x^* \in \cap_{q=0}^{\infty} (-1)^q E_{m_{\ell_q}}$ and the sequence $(m_{\ell_q})_{q \in \mathbb{N}}$ belongs to $\cap_{k=0}^{\infty} \mathscr{P}_k$. So in the application of Theorem I.4.11, we have necessarily

$$\mathscr{P}_\infty(m_\ell)_{\ell \in \mathbb{N}} \subset \bigcap_{k=0}^{\infty} \mathscr{P}_k$$

In other words, this means that this sequence $(m_\ell)_{\ell \in \mathbb{N}}$ is such that, for all subsequences $(m_{\ell_p})_{p \in \mathbb{N}}$, we have

$$\forall k \in \mathbb{N}, \quad \bigcap_{p=0}^{k} (-1)^p E_{m_{\ell_p}} \neq \varnothing$$

Define $(k_\ell)_{\ell \in \mathbb{N}}$ by $k_\ell = m_{2\ell}$ for all $\ell \in \mathbb{N}$. Then, if P and N are two disjoint subsets in $\mathbb{N}$, we can compute

$$\left(\bigcap_{\ell \in N} E_{k_\ell}\right) \cap \left(\bigcap_{\ell \in P} -E_{k_\ell}\right) = \left(\bigcap_{\ell \in N} E_{m_{2\ell}}\right) \cap \left(\bigcap_{\ell \in P} -E_{m_{2\ell}}\right)$$

$$\supset \left(\bigcap_{\ell \in N} E_{m_{2\ell}}\right) \cap \left(\bigcap_{\ell \in N} -E_{m_{2\ell+1}}\right)$$

$$\cap \left(\bigcap_{\ell \in P} -E_{m_{2\ell}}\right) \cap \left(\bigcap_{\ell \in P} E_{m_{2\ell+1}}\right)$$

The sets $2N$, $2N + 1$, $2P$, $2P + 1$ are pairwise disjoint in $\mathbb{N}$, so the set $2N \cup (2N + 1) \cup 2P \cup (2P + 1)$ can be written as a sequence of integers $(p_\ell)_{\ell \in \mathbb{N}}$. Thus we have

$$(\bigcap_{\ell \in N} E_{k_\ell}) \cap (\bigcap_{\ell \in P} -E_{k_\ell}) \supset \sum_{\ell \in \mathbb{N}} (-1)^{p_\ell} E_{m_{p_\ell}} \neq \varnothing$$

This proves that the sequence $(E_{k_\ell}, -E_{k_\ell})_{\ell \in \mathbb{N}}$ is independent in B_{x^*} (see Definition I.4.14).

We are in position to apply Lemma I.4.15, and this will imply that $(x_{k_\ell})_{\ell \in \mathbb{N}}$ is equivalent to the unit vector basis of ℓ_1. The proof of theorem I.4.10 is complete.

An easy but useful consequence of this result is the following.

Corollary I.4.16. If X is a weakly sequentially complete Banach space, either X is reflexive or X contains a subspace, isomorphic to ℓ_1.

5. SUBSPACES OF ℓ_p, $1 \leq p < +\infty$, OR c_0

Among Banach spaces the simplest are ℓ_p and c_0. Here we are going to check what can be said of subspaces of these spaces. Surprisingly, they are not as regular as expected (see Theorem I.5.1). However, their structure is well known.

The first result of this section, Theorem I.5.1, is due to A. M. Davie [Da1, Da2]. Here we reproduce the proof of [LT], Volume I, pp. 87–90.

Theorem I.5.1. For $p > 2$, there exists a subspace of ℓ_p without any basis.

We are going to admit (see [LT], Volume I, p. 87) the existence of an infinite matrix $A = (a_{ij})_{i,j \in \mathbb{N}}$ such that

(1) $\forall i \in \mathbb{N},$ $\{j \in \mathbb{N}, a_{ij} \neq 0\}$ is finite

(2) $\sum_{i=0}^{\infty} \max_{j \in \mathbb{N}} |a_{ij}|^r = M_r < +\infty$ for all $r > \frac{2}{3}$

(3) $A^2 = 0$

(4) $\text{Trace}(A) = \sum_{i=0}^{\infty} a_{ii} \neq 0$

Define

$$\begin{cases} \lambda_i = \underset{j \in \mathbb{N}}{\text{Max}} \, |a_{ij}| & \text{for } i \in \mathbb{N} \\[2mm] b_{ij} = \left(\dfrac{\lambda_j}{\lambda_i}\right)^{1/p+1} a_{ij} & \text{for } i,j \in \mathbb{N} \\[2mm] B = (b_{ij})_{i,j \in \mathbb{N}} \end{cases}$$

Then obviously B verifies $B^2 = 0$, Trace $B = $ Trace $A \neq 0$.

Consider $y_i = (b_{i,1}; b_{i,2}, \, . \, . \,)$, the ith line of the matrix B. Then by (2), since $p/p + 1 > \frac{2}{3}$, we get

$$\sum_{i=0}^{\infty} \| y_i \|_p = \sum_{i=0}^{\infty} \left(\sum_{j=0}^{\infty} |b_{ij}|^p \right)^{1/p}$$

(5)
$$= \sum_{i=0}^{\infty} \left(\sum_{j=0}^{\infty} |a_{ij}|^p \lambda_j^{p/p+1} \frac{1}{\lambda_i^{p/p+1}} \right)^{1/p}$$

$$\leq \sum_{i=0}^{\infty} \lambda_i^{p/p+1} \left(\sum_{j=0}^{\infty} \lambda_j^{p/p+1} \right)^{1/p} \leq M_{p/p+1}^2$$

If $(e_i)_{i \in \mathbb{N}}$ denotes the unit vector basis of ℓ_p and $(e_i^*)_{i \in \mathbb{N}}$ its biorthogonal forms and if $y \in \overline{\text{Span}}[y_i, \, i \in \mathbb{N}]$ in ℓ_p, we get, since $B^2 = 0$,

(6)
$$\sum_{i=0}^{\infty} y_i e_i^*(y) = 0$$

Since Trace $B \neq 0$, we also have

(7)
$$\sum_{i=0}^{\infty} e_i^*(y_i) \neq 0$$

Suppose that the subspace $Y = \overline{\text{Span}}[y_i, \, i \in \mathbb{N}]$ of ℓ_p has a basis $(f_n)_{n \in \mathbb{N}}$ with basis constant K and biorthogonal forms $(f_n^*)_{n \in \mathbb{N}}$. As usual, call $(P_n)_{n \in \mathbb{N}}$ the sequence of canonical projections associated with $(f_n)_{n \in \mathbb{N}}$. Then we have

(8)
$$\sum_{i=0}^{\infty} e_i^* \left(P_N(y_i) \right) = \sum_{i=0}^{\infty} e_i^* \left(\sum_{j=0}^{N} f_j^*(y_i) f_j \right)$$

$$= \sum_{j=0}^{N} f_j^* \left(\sum_{i=0}^{\infty} y_i e_i^*(f_j) \right) = 0 \quad \text{by (6)}$$

(9)
$$\left\|\sum_{i=0}^{\infty} e_i^*(P_N(y_i) - y_i)\right) \le \left\|\sum_{i=0}^{i_0} e_i^*(P_N(y_i) - y_i)\right)\right\|$$

$$+ (K + 1) \sum_{i=i_0+1}^{\infty} \| e_i^* \| \| y_i \|$$

Given $\epsilon > 0$, in the second member of (9) we can choose i_0 such that the second term is smaller than ϵ and then we can choose N such that the first term is less than ϵ. This proves that

$$\sum_{i=0}^{\infty} e_i^*(P_N(y_i) - y_i) = 0$$

Thus (8) and (9) imply

$$\sum_{i=0}^{\infty} e_i^*(y_i) = 0$$

which contradicts (7). Thus $Y = \overline{\text{Span}}[y_i, i \in \mathbb{N}] \subset \ell_p$ has no basis.

Remark I.5.2. The first example of a Banach space without any basis is due to P. Enflo [E1]. There are now other examples of such spaces, for instance, certain subspaces of ℓ_p for $1 \le p \le 2$. For a complete survey of this question, see [LT], Volume I.

The two following propositions show very simple but very important geometric properties of ℓ_p and c_0.

Proposition I.5.3. Let X be ℓ_p $(1 \le p < +\infty)$ or c_0 and $(u_n)_{n\in\mathbb{N}}$ a normalized block basis of the unit vector basis $(e_n)_{n\in\mathbb{N}}$. Then
   (i) $(u_n)_{n\in\mathbb{N}} \overset{1}{\sim} (e_n)_{n\in\mathbb{N}}$ and $\overline{\text{Span}}[u_n, n \in \mathbb{N}]$ is isometric to X.
   (ii) There exists a projection P from X onto $\overline{\text{Span}}[u_n, n \in \mathbb{N}]$, of norm 1.

Proof. Let us prove this result in the case $X = \ell_p$. Let $(e_n)_{n\in\mathbb{N}}$ be the unit vector basis of X.

   (i) Assume that for all $j \in \mathbb{N}$

$$u_j = \sum_{i=m_j+1}^{m_j+1} \lambda_i e_i$$

where $(m_j)_{j\in\mathbb{N}}$ is increasing in $\mathbb{N}$ and $\sum_{i=m_j+1}^{m_{j+1}} |\lambda_i|^p = 1$. Then for $(a_j)_{j\in\mathbb{N}} \subset \mathbb{R}^{(\mathbb{N})}$, we get

$$\left\| \sum_{j=0}^{\infty} a_j u_j \right\| = \left(\sum_{j=0}^{\infty} |a_j|^p \sum_{i=m_j+1}^{m_{j+1}} |\lambda_i|^p \right)^{1/p} = \left(\sum_{j=0}^{\infty} |a_j|^p \right)^{1/p}$$

(ii) For all $j \in \mathbb{N}$, let $u_j^* \in \ell^{p'}$ be such that:

$$\begin{cases} u_j^* \in \overline{\text{Span}}[e_i^*, i \in \{m_j + 1, \ldots, m_{j+1}\}] & \text{and} \quad \| u_j^* \|_{p'} = 1 \\ u_j^*(u_j) = 1 \end{cases}$$

The linear operator P defined by $\forall x \in \ell_p$, $P(x) = \sum_{j=0}^{\infty} u_j^*(x)u_j$ is a projection from ℓ_p onto $\overline{\text{Span}}[u_j, j \in \mathbb{N}]$.

 P is of norm 1 because if we denote $u_j = \sum_{i=n_j+1}^{n_{j+1}} \mu_i e_i^*$, then $\sum_{i=n_j+1}^{n_{j+1}} |\mu_i|^{p'} = 1$ and we can write

$$\| Px \|^p = \sum_{j=0}^{\infty} | u_j^*(x)|^p = \sum_{j=0}^{\infty} \left| \sum_{i=n_j+1}^{n_{j+1}} \mu_i a_i \right|^p$$

$$\leq \sum_{j=0}^{\infty} \left(\sum_{i=n_j+1}^{n_{j+1}} | \mu_i |^{p'} \right)^{p/p'} \sum_{i=n_j+1}^{n_{j+1}} | a_i |^p$$

$$= \sum_{n=0}^{\infty} | a |^p$$

Thus $\| P \| = 1$.

Proposition I.5.4

1. For all $\epsilon > 0$, every infinite-dimensional subspace X of ℓ_p $(1 \leq p < +\infty)$ or c_0 contains a complemented subspace $(1 + \epsilon)$-isomorphic to the whole space.

2. Every complemented infinite-dimensional subspace of ℓ_p $(1 \leq p < +\infty)$ or c_0 is isomorphic to the whole space.

Proof.

1. Let $(x_n)_{n\in\mathbb{N}}$ be a sequence in X that converges to 0 for $\sigma(\ell_p, \ell_{p'},)$ and not for the norm of ℓ_p. Then, by extracting a subsequence and scaling, we can suppose that $\| x_n \| = 1$ for all $n \in \mathbb{N}$. Fix $\epsilon > 0$. Then, as in Proposition I.1.17, there exist a subsequence $(x_{n_k})_{k\in\mathbb{N}}$ of $(x_n)_{n\in\mathbb{N}}$ and a normalized block-basis of the unit vector basis of ℓ_p, $(u_n)_{n\in\mathbb{N}}$, such that

$$\forall k \in \mathbb{N}, \qquad \| x_{n_k} - u_k \|^{p'} \leq \frac{\epsilon^{p'}}{2^{k+1}}$$

Then for all $(a_i)_{i \in \mathbb{N}}$ in $\mathbb{R}^{(\mathbb{N})}$:

$$\left| \left\| \sum_{i=0}^{\infty} a_i x_{n_i} \right\| - \left\| \sum_{i=0}^{\infty} a_i u_i \right\| \right| \leq \left\| \sum_{i=0}^{\infty} a_i (x_{n_i} - u_i) \right\|$$

$$\leq \left(\sum_{i=0}^{\infty} |a_i|^p \right)^{1/p} \left(\sum_{i=0}^{\infty} \| x_{n_i} - u_i \|^{p'} \right)^{1/p'}$$

$$\leq \epsilon \left(\sum_{i=0}^{\infty} |a_i|^p \right)^{1/p}$$

Thus $(x_{n_k})_{k \in \mathbb{N}}$ is $(1 + \epsilon)$-equivalent to $(u_k)_{k \in \mathbb{N}}$ and then $\overline{\text{Span}}[x_{n_k}, k \in \mathbb{N}]$ is $(1 + \epsilon)$-isomorpic to $\overline{\text{Span}}[u_k, k \in \mathbb{N}]$ (see Definition I.1.14 and the following remark), which is isometric to ℓ_p (see Proposition I.5.3(i)). If $(u_k^*)_{k \in \mathbb{N}}$ denote the biothogonal functionals of $(u_k)_{k \in \mathbb{N}}$, then the map P from ℓ_p onto $\overline{\text{Span}}[x_{n_k}, k \in \mathbb{N}]$ such that for x in ℓ_p, $Px = \sum_{k=0}^{\infty} u_k^*(x) x_{n_k}$ is a projection of norm less than $(1 + \epsilon)$, and $\overline{\text{Span}}[x_{n_k}, k \in \mathbb{N}]$ is complemented in ℓ_p (see Proposition I.5.3(ii)).

2. We are going to use *Pelczynski's decomposition method* [Pe2]. Set $X = \ell^p$ $(1 \leq p < +\infty)$ or c_0. Note first that $(X \oplus X) \sim X$. Let Y be a complemented subspace of X. Then $X = Y \oplus X_1$. But if we apply 1 we get that $Y = Z \oplus Y_1$ where $Z \sim X$. Then:

$$X \oplus Y \sim X \oplus (Z \oplus Y_1)$$

$$\sim Z \oplus Z \oplus Y_1$$

$$\sim Z \oplus Y_1$$

$$\sim Y$$

On the other hand, since $(X \oplus X \oplus \cdots)_x = X$, we get

$$X \oplus Y \sim (X \oplus X \oplus \cdots)_X \oplus Y$$

$$= [(Y \oplus X_1) \oplus (Y \oplus X_1) \oplus \cdots]_X \oplus Y$$

$$\sim (X_1 \oplus X_1 \oplus \cdots)_X \oplus (Y \oplus Y \oplus \cdots)_X \oplus Y$$

$$\sim (X_1 \oplus X_1 \oplus \cdots)_X \oplus (Y \oplus Y \oplus \cdots)_X$$

$$\sim [(X_1 \oplus Y) \oplus (X_1 \oplus Y) \oplus \cdots]_X \sim X$$

Both results together give $Y \sim X$.

The next result characterizes in a strong sense the unit vector basis of ℓ_p or c_0. This is the remarkable discovery of M. Zippin concerning perfectly homogenous bases, [Z1]:

Theorem I.5.5. Every Banach space X that has a basis $(e_n)_{n\in\mathbb{N}}$ that is equivalent to all its normalized block-bases is isomorphic to ℓ_p ($1 \leq p < +\infty$) or to c_0.

Proof. First of all, remark that $(e_n)_{n\in\mathbb{N}}$ is in particular equivalent to $(\epsilon_n e_n)_{n\in\mathbb{N}}$ where $\epsilon_n = \pm 1$ for all $n \in \mathbb{N}$. This means that $(e_n)_{n\in\mathbb{N}}$ is unconditional. For any block-basis $U = (u_j)_{j\in\mathbb{N}}$, define T_U by

$$\text{If } x = \sum_{j=0}^{\infty} a_j e_j, \, T_U(x) = T_U\left(\sum_{j=0}^{\infty} a_j e_j\right) = \sum_{j=0}^{\infty} a_j u_j$$

By hypothesis, T_U is bounded. Moreover, if $x = \sum_{j=0}^{\infty} a_j e_j$ is fixed, we have that $\text{Sup}_{U\in\mathcal{U}} \| T_U(x)\| < +\infty$, where $\mathcal{U}$ is the set of all normalized block-bases. Indeed, if not, then there exist $x \in X$ and a normalized block-basis $U^0 = (u_j^0)_{j\in\mathbb{N}}$ such that $\| T_{U^0}(x)\| \geq 2$.

By truncating, we can suppose that there exists N_0 in $\mathbb{N}$ such that

$$\bigcup_{u_j^0 \in U^0} \text{Supp } u_j^0 \leq N_0 < +\infty$$

Then there exists $V^1 \in \mathcal{U}$ such that $\| T_{V^1}(x)\| \geq 4 + \sum_{j=0}^{N_0+1} | a_j |$. If we call U^1 the elements of V^1 whose supports do not hit $[0,1]$, we obviously have $\| T_{U^1}(x)\| \geq 4$.

We can then suppose that there exists N_1 in $\mathbb{N}$ such that

$$\bigcup_{u_j^1 \in U^1} \text{Supp } u_j^1 \leq N_1 < +\infty$$

If we continue this construction, we get a sequence $U^0, U^1, \ldots, U^k,$ $\ldots$ of finite and disjoint normalized block-bases of $(e_n)_{n\in\mathbb{N}}$; the union $U = \bigcup_{i=0}^{\infty} U^i$ is then a normalized block-basis of $(e_n)_{n\in\mathbb{N}}$ such that $\| T_U(x)\| = +\infty$, which contradicts the hypothesis.

So we can apply the Banach-Steinhaus theorem and we get that $\text{Sup}_{U\in\mathcal{U}} \| T_U \| < +\infty$. The same argument would prove that $\text{Sup}_{U\in\mathcal{U}} \| T_U^{-1} \| < +\infty$. This means that there exists a constant M such that, for all $x \in X$,

$$M^{-1} \| x \| \leq \| T_U(x)\| \leq M \| x \|$$

And we can say that $(e_n)_{n\in\mathbb{N}}$ is uniformly equivalent to all its normalized block-bases.

Define $\lambda(n)$ for $n \in \mathbb{N}$ by $\lambda(n) = \| \sum_{i=0}^{n} e_i \|$. Then applying the above inequality to the vectors $x = \sum_{j=0}^{n} e_j$ and

$$u_j = \frac{\sum_{i=(j-1)n^{k-1}+1}^{jn^{k-1}} e_i}{\| \sum_{i=(j-1)n^{k-1}+1}^{jn^{k-1}} e_i \|}, \qquad 1 \le j \le n^k$$

for fixed n and k in $\mathbb{N}$, we get

$$M^{-2}\lambda(n^{k-1})\lambda(n) \le \lambda(n^k) \le M^2(n^{k-1})\lambda(n)$$

So by induction we get, for all n and k,

$$M^{-2k}\lambda(n)^k \le \lambda(n^k) \le M^{2k}\lambda(n)^k$$

Changing k to $k \log m$, we obtain

$$M^{-2k \log m}\lambda(n)^{k \log m} \le \lambda(n^{k \log m}) \le \lambda(m^{k \log n}) \le M^{2k \log n}\lambda(m)^{k \log n}$$

Thus, when $k \to +\infty$ we get, taking logarithms and dividing the extreme members by k,

$$\frac{\log \lambda(n)}{\log n} - \frac{\log \lambda(m)}{\log m} \le 2 \log M \left(\frac{1}{\log n} + \frac{1}{\log m} \right)$$

This proves that the sequence $(\log \lambda(n)/\log n)_{n\in\mathbb{N}}$ is a Cauchy sequence in $\mathbb{R}$, thus converges to some C. In the above inequality, let $m \to +\infty$; then

$$\frac{\log \lambda(n)}{\log n} - C \le 2 \frac{\log M}{\log n}$$

that is, $\lambda(n)/n^C \le M^2$. If we had made $n \to +\infty$ before, we would get in the same way $\lambda(m)/m^C \ge M^{-2}$. And so

$$M^{-2}n^C \le \lambda(n) \le M^2 n^C \quad \text{for all } n \in \mathbb{N}$$

If $C = 0$: Then, for all $(a_i)_{i\in\mathbb{N}} \subset \mathbb{R}^{(\mathbb{N})}$:

$$\left\| \sum_{i=0}^{\infty} a_i e_i \right\| \le M \operatorname*{Sup}_{i\in\mathbb{N}} | a_i | \left\| \sum_{i=0}^{\infty} e_i \right\| \le M^3 \operatorname*{Sup}_{i\in\mathbb{N}} | a_i |$$

$$\left\| \sum_{i=0}^{\infty} a_i e_i \right\| \ge M^{-1} \operatorname*{Sup}_{i\in\mathbb{N}} | a_i |$$

Then $X \sim c_0$.

If $C \neq 0$: Set $p = 1/C$. Assume that $(a_j)_{j=0,\ldots,m} \in \mathbb{R}^{m+1}$ is given by $a_j = k_j/k$, $k_j \in \mathbb{N}$, $k \in \mathbb{N}^*$. Then

$$\left\| \sum_{j=0}^{m} a_j^{1/p} e_j \right\| = k^{-1/p} \left\| \sum_{j=0}^{m} k_j^{1/p} e_j \right\| \geq M^{-3} k^{-1/p} \left\| \sum_{j=0}^{m} \lambda(k_j) e_j \right\|$$

$$\geq M^{-4} k^{-1/p} \left\| \sum_{j=0}^{k_1 + \cdots + k_m} e_j \right\| \geq M^{-4} k^{-1/p} \lambda(k_0 + \cdots + k_m)$$

$$\geq M^{-5} k^{-1/p} \left(\sum_{j=0}^{m} k_j \right)^{1/p} = M^{-5} \left(\sum_{j=0}^{m} a_j \right)^{1/p}$$

The other inequality is obtained by the same method. This gives the equivalence with the ℓ_p-norm by unconditionality and density and this finishes the proof of Theorem I.5.5.

6. REFLEXIVE SPACES AND JAMES SEQUENCE

In this section we give a characterization of reflexive spaces that does not require any knowledge of the dual space. It is due to R. C. James [J4]. See also [Be3].

Theorem I.6.1. Let X be a Banach space; then the following properties are equivalent:

(i) X is not reflexive.

(ii) $\forall \theta \in]0,1[$, $\exists (x_n)_{n \in \mathbb{N}} \subset X$, $\| x_n \| = 1$ for all $n \in \mathbb{N}$, $\exists (f_n)_{n \in \mathbb{N}} \subset X^*$, $\| f_n \| = 1$ for all $n \in \mathbb{N}$, such that

$$f_n(x_k) = \theta \quad \text{if } n \leq k$$

$$= 0 \quad \text{if } n > k$$

(iii) $\forall \theta \in]0,1[$, $\exists (x_n)_{n \in \mathbb{N}} \subset X$, $\| x_n \| = 1$ for all $n \in \mathbb{N}$, such that for all $1 \leq k \leq K$ and $(\alpha_0, \ldots, \alpha_K) \in \mathbb{R}^{K+1}$ with $\sum_{i=0}^{k} \alpha_i = \sum_{i=k+1}^{K} \alpha_i = 1$, then

$$\left\| \sum_{i=0}^{k} \alpha_i x_i - \sum_{i=k+1}^{K} \alpha_i x_i \right\| \geq \theta$$

Note that with condition (iii), called the James condition, we do not need to know X^* to find out whether X is reflexive.

Proof

(ii) $\Rightarrow$ *(iii):* Fix θ in $]0,1[$ and let $(x_n)_{n\in\mathbb{N}}$ and $(f_n)_{n\in\mathbb{N}}$ be given by (ii). With the notation of (iii), we have

$$\left\|\sum_{i=0}^{k}\alpha_i x_i - \sum_{i=k+1}^{K}\alpha_i x_i\right\| \geq \left|f_{k+1}\left(\sum_{i=0}^{k}\alpha_i x_i - \sum_{i=k+1}^{K}\alpha_i x_i\right)\right|$$

$$\geq \theta\left|\sum_{i=k+1}^{K}\alpha_i\right| = \theta$$

(iii) $\Rightarrow$ *(i):* Fix θ in $]0,1[$ and let $(x_n)_{n\in\mathbb{N}}$ be given by (iii). If X is reflexive, denote by A_k the closed convex hull of $(x_n)_{n\geq k}$. Then $(A_k)_{k\in\mathbb{N}}$ is a decreasing sequence of $\sigma(X,X^*)$-compact sets. Then there exists a vector y in $\cap_{k=0}^{\infty}A_k$.

Since $y \in A_0$, there exist $k \in \mathbb{N}$ and $\alpha_0, \ldots, \alpha_k \in \mathbb{R}^{k+1}$ with $\sum_{i=0}^{k}\alpha_i = 1$, such that

$$\left\|y - \sum_{i=0}^{k}\alpha_i x_i\right\| \leq \theta/3$$

Since $y \in A_{k+1}$, there exist $K \in \mathbb{N}$ and $\alpha_{k+1}, \ldots, \alpha_K \in \mathbb{R}^{K-k}$ with $\sum_{i=k+1}^{K}\alpha_i = 1$ such that

$$\left\|y - \sum_{i=k+1}^{K}\alpha_i x_i\right\| \leq \theta/3$$

So we get

$$\left\|\sum_{i=0}^{k}\alpha_i x_i - \sum_{i=k+1}^{K}\alpha_i x_i\right\| \leq \frac{2\theta}{3} < \theta$$

which contradicts (iii).

(i) $\Rightarrow$ *(ii):* We need a lemma.

Lemma I.6.2 (Helly's condition). Let $(f_0, \ldots, f_n) \subset X^*$, $(c_0, \ldots, c_n)$ $\in \mathbb{R}^{n+1}$, $M > 0$ be given. Then the following conditions are equivalent:

(i) For all $\epsilon > 0$, there exists $x \in X$ with $\|x\| \leq M + \epsilon$ such that for all $k = 0, 1, \ldots n$, $f_k(x) = c_k$.

(ii) For all $(a_0, \ldots, a_n)$ in $\mathbb{R}^{n+1}$, $\left|\sum_{i=0}^{n}a_i c_i\right| \leq M\left\|\sum_{i=0}^{n}a_i f_i\right\|$.

Proof of Lemma I.6.2.   If (i) is true, we can write, for all $\epsilon > 0$,

$$\left| \sum_{i=0}^{n} a_i c_i \right| = \left| \sum_{i=0}^{n} a_i f_i(x) \right| \leq \| x \| \left\| \sum_{i=0}^{n} a_i f_i \right\| \leq (M + \epsilon) \left\| \sum_{i=0}^{n} a_i f_i \right\|$$

This proves (ii).

Suppose that (ii) is true and consider φ defined by

$$\varphi : E \rightarrow \mathbb{R}^{n+1}$$

$$x \mapsto (f_0(x), \ldots, f_n(x))$$

If we denote by B the set $\{x \in E, \| x \| \leq M + \epsilon\}$ and by c the n-tuple $(c_0, \ldots, c_n)$, we want to prove that $c \in \varphi(B)$.

Since φ is linear, $\varphi(B)$ is convex, so by the Hahn-Banach theorem if $c \notin \varphi(B)$ there exists $(\beta_0, \ldots, \beta_n) \in \mathbb{R}^{n+1}$, $(\beta_0, \ldots, \beta_n) \neq (0, \ldots, 0)$, such that for all x in B,

$$\sum_{i=0}^{n} \beta_i f_i(x) \leq \sum_{i=0}^{n} \beta_i c_i$$

This implies that

$$(M + \epsilon) \left\| \sum_{i=0}^{n} \beta_i f_i \right\| \leq \left| \sum_{i=0}^{n} \beta_i c_i \right|$$

and therefore, by (ii),

$$\sum_{i=0}^{n} \beta_i f_i = 0$$

If $(f_0, \ldots, f_n)$ are independent in X^*, this implies $(\beta_0, \ldots, \beta_n) = (0, \ldots, 0)$, which is a contradiction. So $c \in \varphi(B)$ and (i) is realized.

If not, we extract an independent subfamily $(f_{i_0}, \ldots, f_{i_k})$ and then, applying the preceding result, there exists $x \in X$, $\| x \| \leq M + \epsilon$, such that

$$f_{i_\ell}(x) = c_{i_\ell} \quad \text{for } \ell = 0, \ldots, k$$

Now if $j \in \{0, \ldots, n\} \backslash \{i_0, \ldots, i_k\}$, there exist $\lambda_0^j, \ldots, \lambda_k^j$ such that

$$f_j = \sum_{\ell=0}^{k} \lambda_\ell^j f_{i_\ell}$$

So

$$f_j(x) = \sum_{\ell=0}^{k} \lambda_\ell^j f_{i\ell}(x)$$

and thus

$$\left| c_j - \sum_{\ell=0}^{k} \lambda_\ell^j c_{i\ell} \right| \leq M \left\| f_j - \sum_{\ell=0}^{k} \lambda_\ell^j f_{i\ell} \right\| = 0$$

Then

$$f_j(x) = \sum_{\ell=0}^{k} \lambda_\ell^j f_{i\ell}(x) = \sum_{\ell=0}^{k} \lambda_\ell^j c_{i\ell} = c_j$$

and (i) is proved.

We come back to the proof of (i) $\Rightarrow$ (ii) in Theorem I.6.1. Suppose that X is not reflexive; then there exists $F \in X^{***}$ such that

$$\| F \| = 1 \quad \text{and} \quad F \big|_X = 0$$

and for all $0 < \theta < 1$, there exists $\xi \in X^{**}$ such that

$$F(\xi) > \theta \quad \text{and} \quad \| \xi \| < 1$$

In particular, this implies that $\| \xi \| > \theta$.

We are going to prove by induction that there exist $(x_n)_{n \in \mathbb{N}} \subset X$ and $(f_n)_{n \in \mathbb{N}} \subset X^*$ such that

$$(*) \begin{cases} \| x_n \| = \| f_n \| = 1 & \text{for all } n \in \mathbb{N} \\ \xi(f_n) = \theta & \text{for all } n \in \mathbb{N} \\ f_n(x_k) = \theta & \text{if } n \leq k \\ \quad\quad = 0 & \text{if } n > k \end{cases}$$

Note that by the choice of ξ we have

$$d(\xi, X) \geq \underset{x \in X}{\text{Inf}} \| \xi - x \| \geq \underset{x \in X}{\text{Inf}} | F(\xi - x)| \geq | F(\xi)| > \theta$$

Let $\eta = \theta / d(\xi, X)$. Then $\eta < 1$.

Suppose that $f_0, \ldots, f_k$ and $x_0, \ldots, x_k$ have been chosen in X^* and X such that

$$\begin{cases} \| f_i \| = \| x_i \| = 1, & 0 \le i \le k \\ \xi(f_i) = \theta, & 0 \le i \le k \\ f_i(x_j) = \theta & \text{if } i \le j \\ f_i(x_j) = 0 & \text{if } i > j \end{cases}$$

Let us prove that we can choose $f_{k+1} \in X^*$ such that

$$\begin{cases} \| f_{k+1} \| = 1 \\ \xi(f_{k+1}) = \theta \\ f_{k+1}(x_j) = 0 & \text{for } j = 0, \ldots, k. \end{cases}$$

Indeed condition (ii) of Lemma I.6.2. is realized in X^* with

$$x_0, \ldots, x_k, \xi \in X^{**}, \quad c_0 = \cdots = c_k = 0, \quad c_{k+1} = \theta, \quad M = \eta$$

because we can write, for all $(a_i)_{i=0,\ldots,k+1}$,

$$| a_{k+1} | \, \theta \le \theta \, \frac{\| \sum_{i=0}^{k} a_i x_i + a_{k+1}\xi \|}{d(X,\xi)} < \left\| \sum_{i=0}^{k} a_i x_i + a_{k+1}\xi \right\|$$

So, by (i) of Lemma I.6.2, there exists $f'_{k+1} \in X^*$ with $\| f'_{k+1} \| < 1$ such that $\xi(f'_{k+1}) = \theta$ and $f'_{k+1}(x_j) = 0$ for $j = 0, \ldots, k$.

Choose one g in the intersection of $\cap_{i=0}^{k}$ Ker x_i and Ker ξ in X^{**}. Then the function $t \to \| f'_{k+1} + tg \|$ is continuous, of norm strictly less than 1 in $t = 0$, and tends to $+\infty$ when $t \to +\infty$. So there is $t_0 > 0$ such that $\| f'_{k+1} + t_0 g \| = 1$. It is clear that $f_{k+1} = f'_{k+1} + t_0 g$ is suitable.

We now choose $x_{k+1} \in X$ such that

$$\begin{cases} \| x_{k+1} \| = 1 \\ f_i(x_{k+1}) = \theta & \text{if } i \le k + 1 \end{cases}$$

Indeed, the same application of Lemma I.6.2 but in X this time, with $f_0, \ldots, f_{k+1} \in X^*, c_0 = c_1 = \cdots = c_{k+1} = \theta, M = \| \xi \|$, shows that it is possible, because we can write

$$\left\| \sum_{i=0}^{k+1} a_i f_i \right\| \ge \frac{| \xi(\sum_{i=0}^{k+1} a_i f_i) |}{\| \xi \|} = \frac{| \sum_{i=0}^{k+1} a_i \theta |}{\| \xi \|}$$

Then the induction works and (*) can be realized. This proves that (i) $\Rightarrow$ (ii) and this finishes the proof of Theorem I.6.1.

We can now give two other very useful characterizations of reflexivity:

Corollary I.6.3

1. A Banach space X in which every bounded sequence has a $\sigma(X,X^*)$-convergent subsequence is reflexive.
2. X is reflexive if and only if all separable closed subspaces of X are reflexive.

Proof. (1) is true because one can remark that the sequence $(x_n)_{n \in \mathbb{N}}$ given by Theorem I.6.1(ii) cannot have a $\sigma(X,X^*)$-convergent subsequence. (2) is true because (ii) or (iii) of Theorem I.6.1 involves only separable subspaces of the space X.

7. TSIRELSON'S SPACE

This is the first example of a separable Banach space containing no subspace isomorphic to ℓ_p ($1 \le p < +\infty$) or c_0. We are going to construct here the dual of the original space, constructed by B. S. Tsirelson [T]. It has also been studied in particular by P. G. Casazza, T. Figiel, W. B. Johnson, and L. Tsafriri in [CJT], [Ca], and [FJ]. These results are also contained in [CS]. See also [BeL] and [LT], Volume I.

Definition I.7.1. A finite set $\{A_0, \ldots ,A_k\}$ of subsets of $\mathbb{N}$ is called *admissible* if

$$k < \text{Min } A_0 < \text{Max } A_0 < \text{Min } A_1 < \text{Max } A_1 < \cdots < \text{Max } A_k$$

We denote by $\mathscr{A}$ the family of all admissible sets.

If $x = (x(n))_{n \in \mathbb{N}} \in \mathbb{R}^{(\mathbb{N})}$ and $A \subset \mathbb{N}$, denote by $P_A x$ the element of $\mathbb{R}^{(\mathbb{N})}$ such that

$$P_A x(n) = x(n) \quad \text{if } n \in A$$

$$= 0 \qquad \text{if } n \notin A$$

Let us define by induction the norm of T: for $x \in \mathbb{R}^{(\mathbb{N})}$ we put

$$\| x \|^{(0)} = \| x \|_{c_0}$$

$$\| x \|^{(n)} = \text{Max} \left[\| x \|^{(n-1)}, \frac{1}{2} \text{Sup} \left\{ \sum_{j=0}^{k} \| P_{A_j} x \|^{(n-1)} / (A_0, \ldots, A_k) \in \mathcal{A} \right\} \right]$$

$$\vdots$$

$$\| x \|_T = \lim_{n \to +\infty} \| x \|^{(n)}$$

This limit exists because the sequence $(\| x \|^{(n)})_{n \in \mathbb{N}}$ is increasing and majorized by the ℓ_1 norm of x.

Definition I.7.2. The Tsirelson space T is the closure of $\mathbb{R}^{(\mathbb{N})}$ under $\| \|_T$.

Proposition I.7.3

1. If $x \in T$, we have

$$\| x \|_T = \text{Max} \left[\| x \|_{c_0}, \frac{1}{2} \text{Sup} \left\{ \sum_{j=0}^{k} \| P_{A_j} x \|_T / (A_0, \ldots, A_k) \in \mathcal{A} \right\} \right]$$

2. The unit vectors $(e_n)_{n \in \mathbb{N}}$ of $\mathbb{R}^{(\mathbb{N})}$ form a 1-unconditional basis of the space T.

3. If $\{u_0, \ldots, u_n\}$ are $(n + 1)$ disjoint normalized blocks on $(e_n)_{n \in \mathbb{N}}$ and if

$$n < \text{Supp } u_0 < \text{Supp } u_1 < \cdots < \text{Supp } u_n$$

we have

$$\forall (a_0, \ldots, a_n) \in \mathbb{R}^{n+1}, \quad \frac{1}{2} \sum_{i=0}^{n} | a_i | \leq \left\| \sum_{i=0}^{n} a_i u_i \right\|_T \leq \sum_{i=0}^{n} | a_i |$$

Proof. The proof of 1 and 2 is obvious. To prove 3 it is enough to remark that the set $\{A_0 = \text{Supp } u_0, \ldots, A_n = \text{Supp } u_n\}$ is admissible in $\mathbb{N}$.

Corollary I.7.4. T does not contain either ℓ_p $(1 < p < +\infty)$ or c_0.

Proof. If T contained ℓ_p $(1 < p < +\infty)$ or c_0, there would exist a sequence of blocks on $(e_n)_{n \in \mathbb{N}}$, equivalent to the unit vector basis of ℓ_p $(1 < p < +\infty)$ or c_0 (see Corollary I.1.18). This is impossible by 3 of Proposition I.7.3.

The proof of the next result comes from [BeL].

Proposition I.7.5. T does not contain ℓ_1.

Proof. If T contained ℓ_1, by Corollary I.1.18 and Theorem I.4.7, for all $\epsilon > 0$ there is a sequence of normalized blocks $(u_n)_{n \in \mathbb{N}}$ on $(e_n)_{n \in \mathbb{N}}$ which verifies, for all $(a_i)_{i \in \mathbb{N}} \in \mathbb{R}^{(\mathbb{N})}$,

$$(1 - \epsilon) \sum_{i=0}^{\infty} | a_i | \leq \left\| \sum_{i=0}^{\infty} a_i u_i \right\|_T \leq \sum_{i=0}^{\infty} | a_i |$$

Thus for all $r \in \mathbb{N}$, we get in particular

$$(*) \qquad \left\| u_0 + \frac{u_1 + \cdots + u_r}{r} \right\| \geq 2(1 - \epsilon)$$

If we choose $r = 2n_0$, where n_0 is the last element in the support of u_0, we will get a contradiction: On the one hand, we have

$$\left\| u_0 + \frac{u_1 + \cdots + u_r}{r} \right\|_{c_0} \leq 1$$

On the other hand, let $(A_j)_{j=0,\ldots,k}$ be an admissible family of sets. If Min $A_0 > n_0$, we get

$$\frac{1}{2} \sum_{j=0}^{k} \left\| P_{A_j} \left(u_0 + \frac{u_1 + \cdots + u_r}{r} \right) \right\|_T \leq \left\| \frac{u_1 + \cdots + u_r}{r} \right\|_T \leq 1$$

which contradicts $(*)$ when ϵ is smaller than $\frac{1}{2}$.

If Min $A_0 \leq n_0$, then necessarily we have $k \leq n_0 = r/2$. Define

$$\begin{cases} \delta = \{i \geq 0 / \| P_{A_j}(u_i)_T \neq 0 \text{ for at least two distinct values of } j\} \\ \sigma = \{i \geq 0 / \| P_{A_j}(u_i)\|_T \neq 0 \text{ for at most one value of } j\} \end{cases}$$

Then $| \delta | \leq k - 1 \leq n_0 - 1$, and we can write

$$\frac{1}{2} \sum_{j=0}^{k} \left\| P_{A_j} \left(u_0 + \frac{u_1 + \cdots + u_r}{r} \right) \right\|_T$$

$$\leq \frac{1}{2} \sum_{j=0}^{k} \| P_{A_j}(u_0) \| + \frac{1}{4n_0} \sum_{i \in \delta} \sum_{j=0}^{k} \| P_{A_j} u_i \|_T + \frac{1}{4n_0} \sum_{i \in \sigma} \sum_{j=0}^{k} \| P_{A_j} u_i \|_T$$

$$\leq 1 + \frac{| \delta |}{2n_0} + \frac{| \sigma |}{4n_0} \leq 1 + \frac{| \delta |}{2n_0} + \frac{2n_0 - | \delta |}{4n_0} = \frac{3}{2} + \frac{| \delta |}{4n_0}$$

$$\leq \frac{7}{4} - \frac{1}{4n_0} \leq \frac{7}{4}$$

which contradicts ($*$) if ϵ is small enough—precisely, smaller than $\frac{1}{8}$. This proves Proposition I.7.5.

NOTES AND REMARKS

1. Banach spaces with unconditional bases have a natural lattice structure. The results that we presented here extend naturally to lattices, in particular those about c_0, ℓ_1, or reflexive spaces (Theorem I.4.3). See [LT], Volume II for a first approach.
2. There are many characterizations of spaces that do not contain ℓ_1; see, for example, [Do, Go, HM, M3, OR, Ros4, Ros5].
3. Tsirelson space has been studied in many different aspects because it is an important source of examples and counterexamples in the theory of Banach spaces. See [BCLT, Ca, CJT, CLL, CO, CS, FJ].
4. James space, which is constructed in Exercise 5 below, is also a source of examples and counterexamples in Banach spaces. References on this space are [CL, J_1].

EXERCISES

Exercises 1 to 5 are from [Be3], Part 2, Chap. II.

1. If $(e_n)_{n\in\mathbb{N}}$ is a basis of X and $(e_n^*)_{n\in\mathbb{N}}$ are the biorthogonal functionals in X^*, show that

$$\forall x^* \in X^*, \qquad \sigma(X^*,X) - \lim_n \sum_{k=0}^n x^*(e_k)e_k^* = x^*$$

2. Let $(e_n)_{n\in\mathbb{N}}$ be an unconditional basis of X and $(e_n^*)_{n\in\mathbb{N}}$ the biorthogonal sequence in X^*. Let $(z_n)_{n\in\mathbb{N}}$ be a bounded sequence of X such that $\forall k \in \mathbb{N}$, $\lim_{n\to+\infty} e_k^*(z_n) = 0$, but which is not $\sigma(X,X^*)$-convergent to 0. Prove that $(z_n)_{n\in\mathbb{N}}$ has a subsequence equivalent to the unit vector basis of ℓ_1.

3. Let $(h_n)_{n\in\mathbb{N}}$ be the Haar system on $[0,1]$. Set

$$g_0 = h_0, \ldots, g_n(t) = \int_0^t h_n(\omega) \, d\omega, \ldots$$

Prove that $(g_n)_{n\in\mathbb{N}}$ is a basis of $C[0,1]$.

4. Suppose that $(e_n)_{n \in \mathbb{N}}$ is a monotone shrinking basis of X. Let $(P_n)_{n \in \mathbb{N}}$ be the canonical projections associated with it and $(e_n^*)_{n \in \mathbb{N}}$ the biorthogonal functionals.

   (a) Show that

$$\forall x^{**} \in X^{**}, \quad \begin{cases} P_n^{**}(x^{**}) = \displaystyle\sum_{i=0}^{n} x^{**}(e_i^*)e_i & \text{for all } n \in \mathbb{N} \\[2mm] \|x^*\| = \lim_{n \to +\infty} \|P_n^{**}(x^{**})\| = \sup_{n \in \mathbb{N}} \|P_n^{**}(x^{**})\| \end{cases}$$

   (b) If $(a_n)_{n \in \mathbb{N}} \in \mathbb{R}^{\mathbb{N}}$ is such that $\sup_{n \in \mathbb{N}} \|\sum_{i=0}^{n} a_i e_i\| < +\infty$, prove that every $\sigma(X^{**}, X^*)$-limit x^{**} of the sequence $(\sum_{i=0}^{n} a_i e_i)_{n \in \mathbb{N}}$ in X^{**} satisfies $x^{**}(e_i^*) = a_i$ for all $i \in \mathbb{N}$. Deduce from this that $(\sum_{i=0}^{n} a_i e_i)_{n \in \mathbb{N}}$ converges to x^{**} for $\sigma(X^{**}, X^*)$.

   (c) Prove that X^{**} is isometric to the space of sequences $(a_n)_{n \in \mathbb{N}}$ such that $\sup_n \|\sum_{i=0}^{n} a_i e_i\| < +\infty$. [Define the isometry by $x^{**} \to (x^{**}(e_j^*))_{j \in \mathbb{N}}$]

5. The James space J: For $x = (x(k))_{k \in \mathbb{N}} \in \mathbb{R}^{\mathbb{N}}$, define:

$$\|x\|_J = \sup \left\{ \left[\sum_{i=0}^{n} (x(p_{2i}) - x(p_{2i+1}))^2 \right]^{1/2} \Big/ n \in \mathbb{N}, \right.$$

$$\left. p_0 < p_1 \cdots < p_{2n+1} < +\infty \right\}$$

$$J = \{x = (x(k))_{k \in \mathbb{N}} \in \mathbb{R}^{\mathbb{N}} / x(k) \xrightarrow[k \to +\infty]{} 0 \text{ and } \|x\|_J < +\infty\}$$

   (a) Show that $\|x\|_J$ is a norm on J and that J is a Banach space.

   (b) Prove that the unit vector basis of $\mathbb{R}^{\mathbb{N}}$ is a monotone basis of J.

   (c) Considering $s_n = e_0 + e_1 + \cdots + e_{n-1}$, $n \in \mathbb{N}$, show that J is not reflexive.

   (d) Prove that $(e_n)_{n \in \mathbb{N}}$ is shrinking. (If not, there is a sequence of consecutive normalized blocks $(u_k)_{k \in \mathbb{N}}$ on $(e_n)_{n \in \mathbb{N}}$, $\delta > 0$ and $x^* \in J^*$ such that $x^*(u_k) \geq \delta$ for all $k \in \mathbb{N}$. But $\sum_{k=1}^{\infty} u_k/k$ belongs to J.)

   (e) Using Exercise 4, show that J^{**} is the space of $x = (x(k))_{k \in \mathbb{N}} \in \mathbb{R}^{\mathbb{N}}$ such that $\|x\|_J < +\infty$.

   (f) Show that $\|x\|_J < +\infty \Rightarrow \lim_{k \to +\infty} x(k)$ exists

(g) Set $s^{**} = \sigma(J^{**},J^*) - \lim_{n \to +\infty} s_n$ in J^{**}. Prove that $s^{**}(e_j^*)$ $= 1$ for all $j \geq 1$ (where the $(e_j^*)_{j \in \mathbb{N}}$ are the biorthogonal forms of the $(e_n)_{n \in \mathbb{N}}$). Show that J^{**} is spanned by s and J and thus is isomorphic to J.

6. Another proof of Proposition I.1.13: (i) $\Rightarrow$ (ii).

(a) Suppose for instance that the right-hand side inequality does not hold. Prove that there exists a sequence $(a^p)_{p \in \mathbb{N}}$, $a^p = (a_i^p)_{i \in \mathbb{N}} \in \mathbb{R}^{\mathbb{N}}$, such that

$$\left\| \sum_{i=0}^{\infty} a_i^p x_i \right\| \xrightarrow[p \to +\infty]{} 0 \quad \text{and} \quad \left\| \sum_{i=0}^{\infty} a_i^p y_i \right\| \xrightarrow[p \to +\infty]{} +\infty$$

(b) Show that there exist $(N_k)_{k \in \mathbb{N}}$ and $(n_k)_{n \in \mathbb{N}}$, increasing sequences of integers, such that for all $k \in \mathbb{N}$

$$\left\| \sum_{i=N_k+1}^{N_{k+1}} a_i^{n_{k+1}} x_i \right\| \leq \frac{1}{2^k} \quad \text{and} \quad \left\| \sum_{i=N_k+1}^{N_{k+1}} a_i^{n_{k+1}} y_i \right\| \geq 2^k$$

(c) Show that $(x_n)_{n \in \mathbb{N}}$ and $(y_n)_{n \in \mathbb{N}}$ are not equivalent.

7. Let $(x_n)_{n \in \mathbb{N}}$ be a sequence in X such that the series $\sum_{i=0}^{\infty} \epsilon_i x_i$ converges in X for all sequences $(\epsilon_i)_{i \in \mathbb{N}}$ such that $\epsilon_i = \pm 1$ for all $i \in \mathbb{N}$. Show that

$$\text{Sup} \left\{ \left\| \sum_{i=n}^{\infty} \epsilon_i x_i \right\| \Big/ \epsilon_i = \pm 1 \right\} \xrightarrow[n \to +\infty]{} 0$$

8. See [LT], Volume I.) Let X be a Banach space such that X^* contains a subspace, isomorphic to c_0. Show that X has a complemented subspace, isomorphic to ℓ_1 and thus X^* has a subspace isomorphic to ℓ_∞. [Define $T : c_0 \to X^*$ and $S = T^*|_x$. Since X is $\sigma(X^{**},X^*)$-dense in X^{**} and $T^*: \sigma(X^{**},X^*) \to \sigma(\ell_1,c_0)$ is continuous, show that $\exists (x_n)_{n \in \mathbb{N}} \subset X$ such that $(x_n)_{n \in \mathbb{N}}$ is bounded, $S(x_n)(n) = 1$ for all n, and $\sum_{i=0}^{n-1} S(x_n)(i) < 1/n$. Prove that $(Sx_n)_{n \in \mathbb{N}}$ has a subsequence $(Sx_{n_k})_{k \in \mathbb{N}}$, equivalent to the unit vector basis of ℓ_1 and such that $\overline{\text{Span}}[Sx_{n_k}, k \in \mathbb{N}]$ is complemented in ℓ_1. Moreover, show that S is an isomorphism from $\overline{\text{Span}}[x_{n_k}, k \in \mathbb{N}]$ onto $\overline{\text{Span}}[Sx_{n_k}, k \in \mathbb{N}]$.]

9. (See [LT], Volume I.) Show that every separable Banach space X is isometric to a quotient of ℓ_1. [Prove that, if $(x_n)_{n \in \mathbb{N}}$ is a dense set in B_x, the map $(a_n)_{n \in \mathbb{N}} \in \ell_1 \to \sum_{n \in \mathbb{N}} a_n x_n$ is an isometry from a quotient of ℓ_1 onto X.]

10. (See [LT], Volume I.) Show that every separable Banach space X is isometric to a subspace of ℓ_∞. [Show that $C(B_{X^*})$ is isometric to ℓ_∞.]

11. Show that, for all $p \in]1, +\infty[$, a sequence $(x_n)_{n \in \mathbb{N}}$ in ℓ_p converges for $\sigma(\ell_p, \ell_{p'})$ if and only if $(x_n)_{n \in \mathbb{N}}$ is bounded and for all $k \in \mathbb{N}$, $(x_n(k))_{n \in \mathbb{N}}$ converges.

12. (See [Be3], Part 2, Chap. IV.) The space ℓ_1 has the Schur property; that is, every sequence in ℓ_1 which converges to 0 for $\sigma(\ell_1, \ell_\infty)$ is norm-convergent to 0.

13. (See [Be3], Part 2, Chap. IV.) Operators from ℓ_r (or c_0) into ℓ_p $(1 \leq p < r < +\infty)$. Recall that an operator T between two Banach spaces X and Y is compact if $\overline{T(B_X)}$ is compact in Y. Recall also that the limit (in the operator norm topology) of a sequence of compact operators is compact and that if T has a finite-dimensional range, T is compact. The aim is to prove that any operator T from ℓ_r into ℓ_p $(1 \leq p < r < +\infty)$ or from c_0 into ℓ_p is compact.

(a) Let P_n (resp. Q_n) be the projections from ℓ_p onto $\overline{\mathrm{Span}}[(e_k)_{k \leq n}]$. Show that if $T : \ell_r \to \ell_p$ is not compact, there exists $\delta > 0$ such that
$$\operatorname*{Inf}_{n,m \in \mathbb{N}} \{ \|(I - Q_n)T(I - P_m)\| \} \geq \delta$$

(b) Take $\epsilon < \delta$. Using the fact that $(e_n)_{n \in \mathbb{N}}$ is a shrinking basis of ℓ_r, prove that there exist a sequence of normalized blocks $(u_i)_{i \in \mathbb{N}}$ on $(e_n)_{n \in \mathbb{N}}$ in ℓ_r and an increasing sequence of integers $(n_i)_{i \in \mathbb{N}}$ such that if $v_i = Tu_i = \sum_{j=0}^{\infty} \beta_j^i e_j$, we have
$$\left\| \sum_{\substack{j \leq n_i \\ \text{or } j > n_{i+1}}} \beta_j^i e_j \right\| \leq \frac{\epsilon}{2^{i+2}} \quad \text{and} \quad \left\| \sum_{j=n_i+1}^{n_{i+1}} \beta_j^i e_j \right\| \geq \delta - \frac{\epsilon}{2}$$

(c) Deduce from this that ℓ_r is isomorphic to ℓ_p. Show that this is not true; therefore T is compact.

(d) Do the same with $T : c_0 \to \ell_p$.

14. c_0 is not complemented in ℓ_∞.

(a) Show that there exists an uncountable family of sets $(A_i)_{i \in I}$ in $\mathbb{N}$ such that
$$\begin{cases} \forall i \in I, & |A_i| = +\infty \\ \forall i \neq j \in I, & |A_i \cap A_j| < +\infty \end{cases}$$
[Consider $A_r = \{ r_n \in \mathbb{Q} / r_n \to r \}$.]

(b) Let $T : \ell_\infty \to \ell_\infty$ such that Ker $T \supset c_0$.

  (i) $\forall i \in I$, let $x_i \in \ell_\infty$, such that $x_i(k) = 0$ if $k \notin A_i$ and $\| x_i \|_\infty = 1$ and $J \subset I$, such that $| J | < +\infty$. Show that $\forall i \in J$, $\exists (y_i, z_i) \in \ell_\infty$ such that

$$\begin{cases} x_i = y_i + z_i \\ \text{Supp } y_i < +\infty \\ (z_i)_{i \in J} \text{ are disjointly supported} \\ \| y_i \|, \| z_i \| \leq 1 \end{cases}$$

  (ii) Deduce from (i) that $\| \sum_{i \in J} Tx_i \|_\infty \leq \| T \|$.

  (iii) Show that $B_n : \{i \in I / Tx_i(n) \neq 0\}$ is countable and that $i \in I \backslash \bigcup_{n \in \mathbb{N}} B_n \Rightarrow Tx_i = 0$.

  (iv) Deduce from (iii) that there exists $A \subset \mathbb{N}, | A | = +\infty$ such that Ker $T \supset \ell_\infty(A)$.

(c) Prove that there is no linear and continuous projection from ℓ_∞ onto c_0.

II

Ultrapowers and Spreading Models

In this chapter we study the notion of ultrapowers of Banach spaces. This concept comes from logic theory, and its introduction in Banach space theory is due to D. Dacunha-Castelle and J. L. Krivine [DCK1]. As a consequence, we will get the definition of spreading models of Banach spaces; these spaces were first constructed by A. Brunel and L. Sucheston [BS1], independently of the definition of ultrapowers. The version of spreading models presented here is slightly different from the original one but gives the same results.

Ultrapowers and spreading models are of interest because it is easier to find ℓ_p ($1 \leq p < +\infty$) or c_0 in an ultrapower or a spreading model of a space than in the space itself.

In some cases, the existence of ℓ_p ($1 \leq p \leq +\infty$) or c_0 in the space can be deduced from analogous properties in ultrapowers or spreading models of this space.

Some of the results of this chapter, especially those about spreading models, can be found in [BeL].

1. FINITE REPRESENTABILITY AND ULTRAPOWERS

Definition II.1.1. A Banach space Y is *finitely representable* (f.r.) in a Banach space X if for all $\epsilon > 0$ and all subspaces Y_0 of Y with dim $Y_0 < +\infty$, there exists a subspace X_0 of X and an isomorphism T_0 from Y_0 onto X_0 such that $\| T_0 \| \| T_0^{-1} \| \leq 1 + \epsilon$. (In other words, if d is the Banach-Mazur distance defined in the introduction, $d(X_0, Y_0) \leq 1 + \epsilon$.)

This notion was introduced by R. C. James in [J7]. It means that all finite subspaces of Y have copies in X.

Remarks II.1.2

1. It is clear that subspaces of a Banach space X are f.r. in X.
2. Property f.r. is *transitive*; that is,
   If Y is f.r. in X and Z is f.r. in Y, then Z is f.r. in X.

Examples II.1.3

1. For any Banach space X, the bidual X^{**} is f.r. in X; this is an easy consequence of the local reflexivity theorem, due to J. Lindenstrauss and H. Rosenthal [LiR] (see also [LT] volume I).
2. If Y is f.r. in X, then in general Y is not isomorphic to a subspace of X; for instance, take $X = (\ell_1^1 \oplus \ell_1^2 \oplus \cdots \oplus \ell_1^n \oplus \cdots)_2$. Then X is reflexive and thus does not contain ℓ_1, but ℓ_1 is f.r. in X.
3. The same example shows that spaces that are f.r. in a reflexive space do not need to be reflexive.
4. Let us recall a definition: X is *uniformly convex* if for all $\epsilon > 0$ there exists $\delta > 0$ such that, for all $x, y \in X$, with $\| x \| = \| y \| = 1$ and $\| x - y \| \geq \epsilon$, then $\| (x + y)/2 \| \leq 1 + \delta$ (see [Be3] or [LT]). Then, obviously, all spaces that are f.r. in X are uniformly convex.
5. In the same spirit, let us recall that X is said to be of *Rademacher type* p ($1 \leq p \leq 2$) if there exists a constant T_p such that, for all $n \in \mathbb{N}$ and $x_1, x_2, \ldots, x_n$ in X, we have

$$\int_0^1 \left\| \sum_{i=0}^n r_i(t)x_i \right\|_X dt \leq T_p \left(\sum_{i=0}^n \| x_i \|^p \right)^{1/p}$$

where $(r_i)_{i \geq 1}$ is the sequence of Rademacher functions on $[0,1]$ described in Definition I.3.7.

Likewise, X is said to be of *Rademacher cotype* q $(2 \leq q \leq +\infty)$ if there exists $C > 0$ such that

$$C_q \int_0^1 \left\| \sum_{i=0}^n r_i(t)x_i \right\|_X dt \geq \left(\sum_{i=0}^n \| x_i \|^q \right)^{1/q}$$

(These definitions were given by J. Hoffman-Jorgensen in [Ho], and these ideas have been studied by G. Pisier [P4].)

Then, obviously, all spaces that are f.r. in a space of Rademacher type p (resp. of Rademacher cotype q) have the same property.

We will see these ideas again in Chapter IV.

Definition II.1.4. Let P be a property.

1. A Banach space X is said to have *Super P* if every space Y, f.r. in X, has P.
2. P is a *superproperty* if $P = $ Super P.

Examples

1. Uniform convexity is a superproperty.
2. To be of Rademacher type p or of cotype q is a superproperty.
3. Reflexivity is not a superproperty (see Example II.1.3.2).

The next results on ultrapowers of Banach spaces are due to D. Dacunha-Castelle and J. L. Krivine [DCK1]. The notion of ultrapower has also been studied by J. Stern [St2].

Let us recall first that an ultrafilter $\mathcal{U}$ on an ordered set I is said to be *trivial* if it consists of all subsets of I containing a given point $i \in I$.

Definition II.1.5. Let I be an ordered set and X a Banach space. Define

$$\ell^\infty(I,X) = \{(x_i)_{i \in I} \in X^I \text{ such that } \sup_{i \in I} \| x_i \| < +\infty\}$$

If $\mathcal{U}$ is a nontrivial ultrafilter on I, on $\ell^\infty(I,X)$, we define the equivalence relation "$\sim$" by

$$(x_i)_{i\in I} \sim (y_i)_{i\in I} \text{ if } \lim_{i,\mathcal{U}} \| x_i - y_i \| = 0$$

The *ultrapower* $\tilde{X} = X^I/\mathcal{U}$ of X is the quotient of $\ell^\infty(I,X)$ by $\sim$, equipped with the norm $\| \tilde{x} \| = \lim_{i,\mathcal{U}} \| x_i \|$ where $(x_i)_{i\in\mathbb{N}}$ is a representative of $\tilde{x}$ in $\ell\infty(I,X)$. We will write $\tilde{x} = (x_i)_{i\in I}$.

Remarks II.1.6

1. If $(X_i)_{i\in X}$ is a family of Banach spaces, we can, in the same way, define the *ultraproduct* of $(X_i)_{i\in I}$ by the quotient of

$$\ell^\infty(I, \prod_{i\in I} X_i) = \{(x_i)_{i\in I} \in \prod_{i\in I} X_i / \sup_{i\in I} \| x_i \| < +\infty\}$$

   by the same equivalence relation $\sim$. We are not going to study this notion in detail here. Properties of ultraproducts are very similar to properties of ultrapowers and proofs are almost the same.

2. If X is finite dimensional, then any ultrapower of X is isometric to X.

3. If $\tilde{x} \in \tilde{X}$ is such that $\| \tilde{x} \| = a$, then there exist representatives $(x_i)_{i\in I}$ of $\tilde{x}$ in X such that $\| x_i \| = a$ for all $i \in I$. Indeed, take any representative $(y_i)_{i\in I}$ of $\tilde{x}$; since $\lim_{i,\mathcal{U}} \| y_i \| = a$, then $x_i = ay_i/ \| y_i \|$ is also a family of representatives of $\tilde{x}$ and $\| x_i \| = a$ for all $i \in I$.

Proposition II.1.7. With the notation of Definition II.1.5, we have the following.

1. $\tilde{X} = X^I/\mathcal{U}$ is a Banach space.
2. X is isometric to a subspace of $\tilde{X}$.
3. $\tilde{X}$ is f.r. in X.
4. If dim $X = +\infty$, $\tilde{X}$ is not separable.

Remark. In the sequel, we will identify X with its image in $\tilde{X}$.

Proof

1. It is clear that $\tilde{X}$ is a normed vector space. Let us prove that it is complete.
   Let $(\tilde{x}^{(n)})_{n\in\mathbb{N}}$ be a Cauchy sequence in $\tilde{X}$. It is enough to prove

that $(\tilde{x}^{(n)})_{n \in \mathbb{N}}$ has a convergent subsequence. So, taking a subsequence, we can suppose that

$$\forall n \in \mathbb{N}, \quad \| \tilde{x}^{(n)} - \tilde{x}^{(n-1)} \| \leq \frac{1}{2^n}$$

Define

$$\begin{aligned} \tilde{u}^{(0)} &= \tilde{x}^{(0)} \\ \tilde{u}^{(1)} &= \tilde{x}^{(1)} - \tilde{x}^{(0)} \\ &\vdots \\ \tilde{u}^{(n)} &= \tilde{x}^{(n)} - \tilde{x}^{(n-1)} \\ &\vdots \end{aligned}$$

Then

$$\tilde{x}^{(n)} = \sum_{k=0}^{n} \tilde{u}^{(k)} \quad \text{and} \quad \| \tilde{u}^{(k)} \| \leq \frac{1}{2^k} \quad \text{for all } n \text{ and } k$$

We have to prove that the series $\sum_{k=0}^{\infty} \tilde{u}^{(k)}$ converges in $\tilde{X}$.

For all $k \in \mathbb{N}$, choose representatives $(u_i^{(k)})_{i \in I}$ of $\tilde{u}^{(k)}$ such that

$$\| u_i^{(k)} \| \leq \frac{1}{2^k} \quad \text{for all } i \in I$$

Since X is a Banach space, for fixed $i \in I$, $\sum_{k=0}^{\infty} u_i^{(k)}$ converges to some u_i in X. Let $\tilde{u}$ be the equivalence class of $(u_i)_{i \in I}$ in $\tilde{X}$. Then

$$\left\| \sum_{k=0}^{n} \tilde{u}^{(k)} - \tilde{u} \right\|_{\tilde{X}} = \lim_{i, \mathcal{U}} \left\| \sum_{k=0}^{n} u_i^{(k)} - u_i \right\|_{X}$$

$$= \lim_{i, \mathcal{U}} \left\| \sum_{k=n+1}^{\infty} u_i^{(k)} \right\| \leq \lim_{i, \mathcal{U}} \frac{1}{2^{n+1}} = \frac{1}{2^{n+1}}$$

This proves that $\sum_{k=0}^{n} \tilde{u}^{(k)}$ converges to $\tilde{u}$ in $\tilde{X}$ and that $\tilde{X}$ is complete.

2. Let

$$Y = \{ \tilde{x} \in \tilde{X} / \exists x \in X, \tilde{x} = (x_i)_{i \in I} \text{ with } x_i = x \text{ for all } i \in I \}$$

Then it is obvious that Y is isometric to X in $\tilde{X}$.

3. Let $\tilde{X}_0$ be a finite-dimensional subspace of $\tilde{X}$. Assume that

$$\tilde{X}_0 = \overline{\text{Span}} \, [\tilde{z}^0, \ldots, \tilde{z}^m] \quad \text{with} \| \tilde{z}^0 \| = \cdots = \| \tilde{z}^m \| = 1$$

Since $\tilde{X}_0$ is finite dimensional, there exists $\delta > 0$ such that for all $(a_k)_{k=0,\ldots,m}$

$$\delta \sum_{k=0}^{m} |a_k| \leq \left\| \sum_{k=0}^{m} a_k \tilde{z}^k \right\|_{\tilde{X}} \leq \sum_{k=0}^{m} |a_k|$$

For all $k = 0, \ldots, m$, choose representatives $(z_i^k)_{i \in I}$ of $\tilde{z}^k$ in X such that $\| z_i^k \| = 1$ for all $i \in I$.

Fix $\epsilon > 0$ and let $(a^\ell)_{\ell \in \{0,\ldots,L\}}$ be a $\delta\epsilon$-net in the unit ball of ℓ_1^{m+1} (see Definition I.1.8). For fixed $\ell \in \{0, \ldots, L\}$, let $a^\ell = (a_k^\ell)_{k=0,\ldots,m}$ be the coordinates of a^ℓ on the unit vector basis of ℓ_1^{m+1}. For all ℓ in $\{0,1,\ldots,L\}$, $(\sum_{k=0}^{m} a_k^\ell z_i^k)_{i \in I}$ is a representative of $\sum_{k=0}^{m} a_k^\ell \tilde{z}_i^k$. Thus we can choose $i_0 \in I$ such that

$$\forall \ell \in \{0, \ldots, L\}, \quad \left| \left\| \sum_{k=0}^{m} a_k^\ell z_{i_0}^k \right\|_X - \left\| \sum_{k=0}^{m} a_k^\ell \tilde{z}^k \right\|_{\tilde{X}} \right| \leq \delta\epsilon$$

Then, if $\sum_{k=0}^{m} |a_k| = 1$, there exists $\ell \in \{0, \ldots, L\}$ such that

$$\sum_{k=0}^{m} |a_k - a_k^\ell| \leq \delta\epsilon$$

and thus we get

$$\left| \left\| \sum_{k=0}^{m} a_k z_{i_0}^k \right\|_X - \left\| \sum_{k=0}^{m} a_k \tilde{z}^k \right\|_{\tilde{X}} \right|$$

$$\leq \left\| \sum_{k=0}^{m} a_k z_{i_0}^k - \sum_{k=0}^{m} a_k^\ell z_{i_0}^k \right\|_{\tilde{X}}$$

$$+ \left\| \sum_{k=0}^{m} a_k^\ell z_{i_0}^k - \sum_{k=0}^{m} a_k^\ell \tilde{z}^k \right\|_{\tilde{X}} + \left\| \sum_{k=0}^{m} a_k^\ell \tilde{z}^k - \sum_{k=0}^{m} a_k \tilde{z}^k \right\|_{\tilde{X}}$$

$$\leq \delta\epsilon + \sum_{k=0}^{m} |a_k - a_k^\ell| (\| z_{i_0}^k \|_X + \| \tilde{z}^k \|_{\tilde{X}}) \leq 3\delta\epsilon$$

So, if $(a_k)_{k=0,\ldots,m}$ are any real numbers, we get

$$\left| \left\| \sum_{k=0}^{m} a_k z_{i_0}^k \right\|_X - \left\| \sum_{k=0}^{m} a_k \tilde{z}^k \right\|_{\tilde{X}} \right| \leq 3\delta\epsilon \sum_{k=0}^{m} |a_k| \leq 3\epsilon \left\| \sum_{k=0}^{m} a_k \tilde{z}^k \right\|_{\tilde{X}}$$

So the subspace $\overline{\text{Span}} \, [z_{i_0}^k, \, k = 0, \ldots, m]$ of X is at distance less than 3ϵ to $\bar{X}_0$, and this proves 3.

4. Let $(x_n)_{n \in \mathbb{N}}$ be a sequence in X such that $\| x_n \| = 1$ for all $n \in \mathbb{N}$ and $\| x_n - x_m \| \geq 1$ if $n \neq m$. We build such a sequence by an easy induction: if x_1 is chosen, let $x_1^* \in X^*$ such that $x_1^*(x_1) = 1$, $\| x_1^* \| = 1$. Then take $x_2 \in \text{Ker } x_1^*$. Thus $\| x_1 - x_2 \| \geq x_1^*(x_1 - x_2) = 1$, and so on.

Let $(A_\alpha)_{\alpha \in \wedge}$ be an uncountable family of subsets of $\mathbb{N}$ such that $| A_\alpha | = +\infty$ for all $\alpha \in \wedge$ and $| A_\alpha \cap A_\beta | < +\infty$ if $\alpha \neq \beta$. [To construct such a family, index $\mathbb{N}$ by $\mathbb{Q}$ and take $\wedge = \mathbb{R}\backslash\mathbb{Q}$; for all $\alpha \in \mathbb{R}\backslash\mathbb{Q}$, choose a sequence of rational numbers $(r_n^\alpha)_{n \in \mathbb{N}}$ converging to α. Set $A_\alpha = (r_n^\alpha)_{n \in \mathbb{N}}$]. Let $\mathcal{U}$ be a nontrival ultrafilter on $\mathbb{N}$. In $\bar{X} = X^{\mathbb{N}}/\mathcal{U}$, define for all $\alpha \in \wedge$, $\bar{x}_\alpha = (x_{i_\alpha})_{i_\alpha \in A_\alpha}$.

Then $\| \bar{x}_\alpha - \bar{x}_\beta \| \geq 1$ if $\alpha \neq \beta$. Since $\wedge$ is not countable, this proves that $\bar{X}$ is not separable.

The next result is the converse of 3 in Proposition II.1.7. The proof of Theorem II.1.8 that we give here is due to S. Heinrich [H]. This result was proved by J. Stern [St2].

Theorem II.1.8. If Y is finitely representable in a Banach space X, then there exists an ultrapower $\bar{X}$ of X such that Y is isometric to a subspace of $\bar{X}$.

Proof. Denote by I the set of all pairs (M, ϵ) where M is a finite-dimensional subspace of Y and $\epsilon > 0$. We put a partial order on I by the definition

$$(M_1, \epsilon_1) < (M_2, \epsilon_2) \quad \text{if } \begin{cases} M_1 \subset M_2 \\ \epsilon_1 \geq \epsilon_2 \end{cases}$$

Set $\mathcal{F} = \{I_0 \subset I / \exists (M_0, \epsilon_0) \in I$ such that $I_0 = \{(M, \epsilon) > (M_0, \epsilon_0)\}\}$. $\mathcal{F}$ is the natural order filter on I.

Let $\mathcal{U}$ be an ultrafilter, dominating $\mathcal{F}$ on I, and denote by $\bar{X}$ the ultrapower $X^I/\mathcal{U}$ of X. By hypothesis, for all $(M, \epsilon) \in I$, there exist a subspace N of X and a $(1 + \epsilon)$-isomorphism T from M onto N.

Define a linear map $J : Y \to X^I/\mathcal{U}$ as follows: For all x in Y,

$$Jx = (y_{(M, \epsilon)})_{(M, \epsilon) \in I} \quad \text{where } y_{(M, \epsilon)} = \begin{cases} Tx & \text{if } x \in M \\ 0 & \text{if } x \notin M \end{cases}$$

If $\epsilon_0 > 0$ and $x \in Y$ are fixed, define $I_0 = \{(M,\epsilon)/x \in M \text{ and } \epsilon \le \epsilon_0\}$. Then $I_0 \in \mathcal{F}$ and, by definition, there exists $J_0 \in \mathcal{U}$ such that $J_0 \subset I_0$. Thus if $(M,\epsilon) \in J_0$, we get

$$(1 + \epsilon_0)^{-1} \| x \| \le \| y_{(M,\epsilon)} \| \le (1 + \epsilon_0) \| x \|$$

and by definition of $X'/\mathcal{U}$, this implies

$$(1 + \epsilon_0)^{-1} \| x \| \le \| Jx \| \le (1 + \epsilon_0) \| x \|$$

Since this is true for all $\epsilon_0 > 0$, we get that J is an isometry.

2. SPREADING MODELS

This notion was first introduced in by A. Brunel and L. Sucheston in [BS1]. The authors used a version of Ramsey's theorem for this construction. Here we present J. L. Krivine's construction of spreading models, based on the notion of ultrapowers; it is reproduced from [BeL]. We obtain almost the same construction, except that indices are inversely ordered. Since properties of spreading models are the same in both constructions, we will not introduce a new terminology for this construction, as it was done in [BeL]. We use the same notation as in [BeL].

In this part, Banach spaces will always be separable.

Let $\mathcal{U}$ be a *nontrivial* ultrafilter on $\mathbb{N}$ (recall that an ultrafilter is said to be *trivial* if it consists of all subsets of $\mathbb{N}$ which contain a given integer k).

Reiteration of ultrapowers

Let X be a separable Banach space and $(x_n)_{n \in \mathbb{N}}$ be a bounded sequence in X with no Cauchy subsequence. Set

$$\begin{cases} X^0 = X^{\mathbb{N}}/\mathcal{U} \\ e_0 = (x_n)_{n \in \mathbb{N}} \in X^0 \end{cases}$$

Then obviously $e_0 \notin X$.

Let $(x_n^0)_{n \in \mathbb{N}}$ be the images of $(x_n)_{n \in \mathbb{N}}$ by the canonical isometry from X into X^0. Set

$$\begin{cases} X^1 = X^{0\mathbb{N}}/\mathcal{U} \\ e_1 = (x_n^0)_{n \in \mathbb{N}} \in X^1 \end{cases}$$

Then $e_1 \notin X^0$; indeed, if e_1 was in X^0 (that means in the canonical image of X^0 in X^1) there would exist a sequence $(y_n)_{n \in \mathbb{N}}$ in X such that

$$\begin{cases} \varphi_0 = (y_n)_{n \in \mathbb{N}} \in X^0 \\ e_1 = (\varphi_0, \varphi_0, \ldots) \end{cases}$$

Thus, by definition,

$$0 = \lim_{n, \mathcal{U}} \| \varphi_0 - x_n^0 \|_{X^0} = \lim_{n, \mathcal{U}} \lim_{m, \mathcal{U}} \| y_m - x_n \|_X$$

Since $(x_n)_{n \in \mathbb{N}}$ has no Cauchy subsequence, there exists $\delta > 0$ such that

$$\| x_p - x_q \| \geq \delta \quad \text{for all } p, q \in \mathbb{N}, \, p \neq q$$

Thus:

$$\| x_p - y_m \| + \| x_q - y_m \| \geq \| x_p - y_m - x_q + y_m \| \geq \delta$$

which contradicts the above property.

We can iterate this construction and get a sequence of spaces $X \subset X^0 \subset X^1 \subset \cdots \subset X^n \subset \cdots$ and a sequence $(e_0, e_1, \ldots, e_n, \ldots)$ such that for all i, $e_i \in X^i$ and if $j < i$, $e_i \notin X^j$. Define:

$$X^\infty = \overline{\bigcup_{k=0}^{\infty} X^k}$$

Then in X^∞, by construction, we can write for all $x \in X$ and $(a_k)_{k \in \mathbb{N}} \in \mathbb{R}^{(\mathbb{N})}$:

$$(*) \quad \left\| x + \sum_{k=0}^{\infty} a_k e_k \right\|_{X^\infty} = \cdots \lim_{n_k, \mathcal{U}} \lim_{n_{k-1}, \mathcal{U}} \cdots \lim_{n_0, \mathcal{U}} \left\| x + \sum_{k=0}^{\infty} a_k x_{n_k} \right\|_X$$

Remark. In this setting, we are identifying X with its images in X^0, $X^1, \ldots, X^n, \ldots, X^\infty$. From now on, we will always make this identification.

Definition II.2.1

1. $\overline{\text{Span}} \, [e_k, k \in \mathbb{N}]$ in X^∞ is called the *spreading model* associated with $(x_m)_{n \in \mathbb{N}}$ and $\mathcal{U}$.
2. $X + \overline{\text{Span}} \, [e_k, k \in \mathbb{N}]$ in X^∞, where the norm of this sum is defined by $(*)$, is called the *spreading model over X* associated with $(x_n)_{n \in \mathbb{N}}$ and $\mathcal{U}$.
3. $(e_n)_{n \in \mathbb{N}}$ is called the *fundamental sequence* of the spreading model.

Let us give the first properties of fundamental sequences of spreading models.

Proposition II.2.2. Let $(x_n)_{n \in \mathbb{N}}$ be a bounded sequence in a separable Banach space X, with no Cauchy subsequence, and let $(e_n)_{n \in \mathbb{N}}$ be the fundamental sequence of the spreading model associated with $(x_n)_{n \in \mathbb{N}}$ and an ultrafilter $\mathcal{U}$ on $\mathbb{N}$.

1.   For all $x \in X$, $(a_k)_{k \in \mathbb{N}} \in \mathbb{R}^{(\mathbb{N})}$, and all nondecreasing sequence $(n_i)_{i \in \mathbb{N}}$ in $\mathbb{N}$,

$$\left\| x + \sum_{i=0}^{\infty} a_i e_i \right\| = \left\| x + \sum_{i=0}^{\infty} a_i e_{n_i} \right\|$$

2.   The spreading model $\overline{\text{Span}}\,[e_n,\, n \in \mathbb{N}]$ is f.r. in X.

Proof

1.   As in [BeL], let us prove, for example, that

$$\left\| x + a_0 e_0 + a_1 e_1 + \sum_{i=3}^{k} a_i e_i \right\|_{X^\infty} = \left\| x + a_0 e_0 + a_1 e_2 + \sum_{i=3}^{k} a_i e_i \right\|_{X^\infty}$$

By definition, we have

$$\left\| x + a_0 e_0 + a_1 e_1 + \sum_{i=3}^{k} a_i e_i \right\|_{X^\infty}$$

$$= \left\| x + a_0 e_0 + a_1 e_1 + \sum_{i=3}^{k} a_i e_i \right\|_{X^k}$$

$$= \lim_{n_k, \mathcal{U}} \cdots \lim_{n_3, \mathcal{U}} \left\| x + a_0 e_0 + a_1 e_1 + \sum_{i=3}^{k} a_i x_{n_i} \right\|_{X^1}$$

$$= \lim_{n_k, \mathcal{U}} \cdots \lim_{n_3, \mathcal{U}} \lim_{m, \mathcal{U}} \left\| x + a_0 e_0 + a_1 x_m + \sum_{i=3}^{k} a_i x_{n_i} \right\|_{X^0}$$

$$= \lim_{n_k, \mathcal{U}} \cdots \lim_{n_3, \mathcal{U}} \lim_{m, \mathcal{U}} \lim_{p, \mathcal{U}} \left\| x + a_0 x_p + a_1 x_m + \sum_{i=3}^{k} a_i x_{n_i} \right\|_{X}$$

and also

$$\left\| x + a_0 e_0 + a_1 e_2 + \sum_{i=3}^{k} a_i e_i \right\|_{X^\infty}$$

$$= \left\| x + a_0 e_0 + a_1 e_2 + \sum_{i=3}^{k} a_i e_i \right\|_{X^k}$$

$$= \lim_{n_k, \mathcal{U}} \cdots \lim_{n_3, \mathcal{U}} \left\| x + a_0 e_0 + a_1 e_2 + \sum_{i=3}^{k} a_i x_{n_i} \right\|_{X^2}$$

$$= \lim_{n_k, \mathcal{U}} \cdots \lim_{n_3, \mathcal{U}} \lim_{m, \mathcal{U}} \left\| x + a_0 e_0 + a_1 x_m + \sum_{i=3}^{k} a_i x_{n_i} \right\|_{X^0}$$

$$= \lim_{n_k, \mathcal{U}} \cdots \lim_{n_3, \mathcal{U}} \lim_{m, \mathcal{U}} \lim_{p, \mathcal{U}} \left\| x + a_0 x_p + a_1 x_m + \sum_{i=3}^{k} a_i x_{n_i} \right\|_X$$

So these two quantities are equal.
2. is obvious by construction.

We can how give the corresponding definition:

Definition II.2.3

1. A sequence $(e_n)_{n \in \mathbb{N}}$ such that, for all k in $\mathbb{N}$, all $(a_0 \ldots a_k)$ in $\mathbb{R}^{k+1}$, and all increasing sequences of integers $(n_i)_{i \in \mathbb{N}}$, $\left\| \sum_{i=0}^{\infty} a_i e_i \right\| = \left\| \sum_{i=0}^{\infty} a_i e_{n_i} \right\|$, is said to be *invariant by spreading* (*I.S.*). We can also say that a sequence is I.S. if and only if all its subsequences are 1-equivalent to itself.
2. If, moreover, we have

$$\forall x \in X, \quad \left\| x + \sum_{i=0}^{\infty} a_i e_i \right\| = \left\| x + \sum_{i=0}^{\infty} a_i e_{n_i} \right\|$$

then $(e_n)_{n \in \mathbb{N}}$ is said to be *invariant by spreading over X* (*I.S. over X*).

Examples. The unit vector basis of ℓ_p ($1 \leq p < +\infty$) or c_0 is invariant by spreading.

We can say a little more than Proposition II.2.2.2; let us first give a definition:

Definition II.2.4. A sequence $(e_n)_{n \in \mathbb{N}}$ is *block finitely representable* (*block f.r.*) in a sequence $(x_n)_{n \in \mathbb{N}}$ if for all $k \in \mathbb{N}$ and $\epsilon > 0$ there exist

$k + 1$ blocks $u_0, \ldots, u_k$ on $(x_n)_{n\in\mathbb{N}}$ such that

$$d[\text{Span}\,(u_0, \ldots, u_k), \text{Span}\,(e_0, \ldots, e_k)] \leq 1 + \epsilon$$

where d is the Banach-Mazur distance.

Remark. This property is *transitive*; that is, if $(e_n)_{n\in\mathbb{N}}$ is block f.r. in $(x_n)_{n\in\mathbb{N}}$ and $(x_n)_{n\in\mathbb{N}}$ is block f.r. in $(y_n)_{n\in\mathbb{N}}$, then $(e_n)_{n\in\mathbb{N}}$ is block f.r. in $(y_n)_{n\in\mathbb{N}}$.

Proposition II.2.5. Under the hypothesis of Proposition II.2.2, $(e_n)_{n\in\mathbb{N}}$ is block f.r. in $(x_n)_{n\in\mathbb{N}}$.

The proof is obvious by construction, with blocks $u_0 = x_{n_0}, \ldots, u_k = x_{n_k}$ for suitable $n_0 < \cdots < n_k$.

Definition II.2.6

1. Recall that $\lim_{n_0 > n_1 \cdots > n_k}$ means that the limit is taken when the $(k + 1)$-tuple $(n_0, \ldots, n_k)$ tends to ∞ in the order $n_0 > n_1 \cdots > n_k$.
2. A sequence $(x_n)_{n\in\mathbb{N}}$ such that for all $k \in \mathbb{N}$, $(a_0 \ldots a_k)$ in $\mathbb{R}$, and x in X, $\lim_{n_0 > n_1 > \cdots > n_k} \| x + a_0 x_{n_0} + a_1 x_{n_1} + \cdots + a_k x_{n_k} \|$ exists is called a *good sequence*.

The following result is called *extraction of good subsequences*. It is due to A. Brunel and L. Sucheston [BS1] and is also given in [BeL]:

Proposition II.2.7. Given a bounded sequence $(x_n)_{n\in\mathbb{N}}$ with no Cauchy subsequence in a separable Banach space X and a nontrivial ultrafilter $\mathcal{U}$ on $\mathbb{N}$, there exists a subsequence $(x'_n)_{n\in\mathbb{N}}$ of $(x_n)_{n\in\mathbb{N}}$ such that if $(e_n)_{n\in\mathbb{N}}$ is the fundamental sequence of the spreading model associated with $(x_n)_{n\in\mathbb{N}}$ and $\mathcal{U}$, we get: For all k in $\mathbb{N}$, all $(a_0, \ldots, a_k)$ in $\mathbb{R}$, and all x in X,

$$\| x + a_0 e_0 + \cdots + a_k e_k \| = \lim_{n_0 > n_1 \cdots > n_k} \| x + a_0 x'_{n_0} + \cdots + a_k x'_{n_k} \|$$

$(x'_n)_{n\in\mathbb{N}}$ is called a *good subsequence* of $(x_n)_{n\in\mathbb{N}}$.

Proof. By approximation it is clear that it is enough to prove this result for x in a countable dense set in X and $(a_0, a_1, \ldots, a_k)$ in $\mathbb{Q}^{(\mathbb{N})}$

instead of X and $\mathbb{R}^{(\mathbb{N})}$. Then, by a diagonal process, it is enough to prove it for fixed x in X, k in $\mathbb{N}$, and $(a_0, \ldots, a_k)$ in $\mathbb{R}^{k+1}$.

Suppose for simplicity that $k = 1$. Then, by definition, we can write that for all $\epsilon > 0$ there exists A^ϵ in $\mathcal{U}$ such that

$$\forall n \in A^\epsilon, \qquad \left| \, \| x + a_0 e_0 + a_1 e_1 \| - \| x + a_0 e_0 + a_1 x_n \| \, \right| \leq \epsilon$$

Then, for all n in A^ϵ, there exists B_n^ϵ in $\mathcal{U}$ such that

$$\forall m \in B_n^\epsilon \qquad \left| \, \| x + a_0 e_0 + a_1 x_n \| - \| x + a_0 x_m + a_1 x_n \| \, \right| \leq \epsilon$$

Since $\mathcal{U}$ is an ultrafilter, by induction we can choose

$$\begin{cases} k_0 \in A^{1/2} \\ k_1 \in A^{1/4} \cap B_{k_0}^{1/4} \\ k_2 \in A^{1/2^3} \cap B_{k_0}^{1/2^3} \cap B_{k_1}^{1/2^3} \\ \quad \vdots \\ k_n \in A^{1/2^{n+1}} \cap B_{k_0}^{1/2^{n+1}} \cap \cdots \cap B_{k_{n-1}}^{1/2^{n+1}} \\ \quad \vdots \end{cases}$$

It is easy to see that, for this subsequence $(x'_n)_{n \in \mathbb{N}} = (x_{k_n})_{n \in \mathbb{N}}$, we get

$$\| x + a_0 e_0 + a_1 e_1 \| = \lim_{n_0 > n_1} \| x + a_0 x'_{n_0} + a_1 x'_{n_1} \|$$

Remark II.2.8. In general, a given bounded sequence in X with no Cauchy subsequence can have more than one spreading model; it depends on the ultrafilter $\mathcal{U}$. On the contrary, if $(x_n)_{n \in \mathbb{N}}$ is a good sequence, then it has a unique spreading model; indeed, if $\lim_{n_0 > \cdots > n_k} \| x + a_0 x_{n_0} + \cdots + a_k x_{n_k} \|$ exists, then

$$\lim_{n_0} \cdots \lim_{n_k} \| x + a_0 x_{n_0} + \cdots + a_k x_{n_k} \|$$

exists too and any ultrafilter $\mathcal{U}$ will give the same limit.

3. PROPERTIES OF I.S. SEQUENCES

The behavior of a good sequence $(x_n)_{n \in \mathbb{N}}$ and the associated I.S sequence $(e_n)_{n \in \mathbb{N}}$ was studied by J. T. Lapresté and the author in [GL1]. These results are also written in [BeL].

Let us start with the spreading model:

Proposition II.3.1.   Let $(e_n)_{n\in\mathbb{N}}$ be a I.S. sequence in a separable Banach space Y; then one and only one of these properties is realized:

(i) $(e_n) \to_{n\to+\infty} e$, $e \neq 0$ for $\sigma(Y, Y^*)$.

(ii) $(e_n) \to_{n\to+\infty} 0$ for $\sigma(Y, Y^*)$.

(iii) (e_n) is $\sigma(Y, Y^*)$-Cauchy, non-$\sigma(Y, Y^*)$-convergent.

(iv) $(e_n)_{n\in\mathbb{N}}$ is equivalent to the unit vector basis of ℓ_1.

Proof.   Since $(e_n)_{n\in\mathbb{N}}$ is I.S., it is bounded. Moreover, since $(e_n)_{n\in\mathbb{N}}$ is 1-equivalent to all its subsequences, this result is a direct consequence of Rosenthal's theorem (I.4.10).

Corollary II.3.2.   With the notation of Proposition II.3.1:

(a)   In (ii), (iii), (iv), $(e_n)_{n\in\mathbb{N}}$ is basic.

(b)   In (i), we get, for all $(a_n)_{n\in\mathbb{N}}$ in $\mathbb{R}^{(\mathbb{N})}$,

$$\left\| \sum_{i=0}^{\infty} a_i e_i \right\| \geq \| e \| \left| \sum_{i=0}^{\infty} a_i \right|$$

(c)   In (iii), if e'' denotes the $\sigma(Y^{**}, Y^*)$-limit of $(e_n)_{n\in\mathbb{N}}$, we get, for all $(a_n)_{n\in\mathbb{N}}$ in $\mathbb{R}^{(\mathbb{N})}$,

$$\left\| \sum_{i=0}^{\infty} a_i e_i \right\| \geq \| e'' \| \left| \sum_{i=0}^{\infty} a_i \right|$$

and

$$\left\| \sum_{i=0}^{\infty} a_i e_i \right\| \geq \frac{\| e'' \|}{M + 1} \sup_{q\in\mathbb{N}} \left| \sum_{i\geq q} a_i \right|$$

(where M is the basic constant of $(e_n)_{n\in\mathbb{N}}$).

(d)   In (ii), $(e_n)_{n\in\mathbb{N}}$ is basic and 1-unconditional.

Proof.   (a) is a consequence of Theorem I.1.10 and the fact that $(e_n)_{n\in\mathbb{N}}$ is 1-equivalent to all its subsequences.

(b)   For $(a_0, \ldots, a_k) \in \mathbb{R}^{k+1}$, we can write

$$\left\| \sum_{i=0}^{k} a_i e_i \right\| = \left\| \sum_{i=0}^{k} a_i e_{n_i} \right\| \quad \text{for all } n_0 < n_1 < \cdots n_k$$

$$= \lim_{n_k \to +\infty} \left\| \sum_{i=0}^{k} a_i e_{n_i} \right\|$$

$$\geq \left\| \sum_{i=0}^{k-1} a_i e_{n_i} + a_k e \right\| \geq \cdots \geq \| e \| \left| \sum_{i=0}^{k} a_i \right|$$

(c)   The first part is similar to (b) with e'' instead of e and the second part follows from the obvious property

$$\left\| \sum_{i=q}^{k} a_i e_i \right\| \leq (M + 1) \left\| \sum_{i=0}^{k} a_i e_i \right\|$$

(d)   Let us prove that, for $a_{i_0} \neq 0$,

$$\left\| \sum_{\substack{i=0 \\ i \neq i_0}}^{k} a_i e_i \right\| \leq \left\| \sum_{i=0}^{k} a_i e_i \right\|$$

Since $(e_n) \to_{n \to +\infty} 0$ for $\sigma(Y, Y^*)$, then 0 belongs to the $\sigma(Y, Y^*)$-closed convex hull of $(e_n)_{n \in \mathbb{N}}$. Thus it is also in the $\| \ \|$-closed convex hull of $(e_n)_{n \in \mathbb{N}}$. So there exists N and $p_0, \ldots, p_k$ in $\mathbb{N}$ such that

$$\begin{cases} \sum_{i=0}^{\ell} p_i = N \\ \dfrac{1}{N} \left\| \sum_{i=0}^{\ell} p_i e_{i+j} \right\| \leq \epsilon / |a_{i_0}| \quad \text{for all } j \in \mathbb{N} \end{cases}$$

Note that $E_j = \sum_{i=0}^{i_0-1} a_i e_i + a_{i_0} e_{i_0+j} + \sum_{i=i_0+1}^{k} a_{i_0} e_{i+\ell}$ for $0 \leq j \leq \ell$ and observe that, because $(e_n)_{n \in \mathbb{N}}$ is I.S., we have $\| E_j \| = \| \sum_{i=0}^{k} a_i e_i \|$. Then

$$\left\| \sum_{j=0}^{\ell} p_j E_j \right\| = N \left\| \sum_{i=0}^{i_0-1} a_i e_i + a_{i_0} \frac{1}{N} \sum_{j=0}^{\ell} p_j e_{i_0+j} + \sum_{i=i_0+1}^{k} a_i e_{i+\ell} \right\|$$

$$\leq \sum_{j=0}^{\ell} p_j \| E_j \| \leq N \left\| \sum_{i=0}^{k} a_i e_i \right\|$$

On the other hand, we can write

$$\left\| \sum_{j=0}^{\ell} p_j E_j \right\| \geq N \left(\left\| \sum_{\substack{i=0 \\ i \neq i_0}}^{k} a_i e_i \right\| - \epsilon \right)$$

When $\epsilon \to 0$, this proves the corollary.

The next result comes from [BS2].

Proposition II.3.3.   Let $(e_n)_{n \in \mathbb{N}}$ be an I.S. sequence in a separable space X. Then $(e_{2n} - e_{2n+1})_{n \in \mathbb{N}}$ is 1-unconditional.

Proof. It is enough to show that for a given finite sequence $(a_n)_{n \in \mathbb{N}}$ in $\mathbb{R}$:

$$\left\| \sum_{i \neq i_0} a_i(e_{2i} - e_{2i+1}) \right\| \leq \left\| \sum_{i=0}^{\infty} a_i(e_{2i} - e_{2i+1}) \right\|$$

We can write, for $k \leq m$,

$$\left\| \sum_{i=0}^{\infty} a_i(e_{2i} - e_{2i+1}) \right\| = \left\| \sum_{i=0}^{i_0-1} a_i(e_{2i} - e_{2i+1}) \right.$$
$$+ a_{i_0}(e_{2(i_0+k)} - e_{2(i_0+k)+1})$$
$$\left. + \sum_{i=i_0+m+1}^{\infty} a_i(e_{2i} - e_{2i+1}) \right\|$$

Summing these equalities for $1 \leq k \leq m$, we find

$$\left\| \sum_{i=0}^{\infty} a_i(e_{2i} - e_{2i+1}) \right\| \geq \left\| \sum_{i=0}^{i_0-1} a_i(e_{2i} - e_{2i+1}) + \frac{a_{i_0}}{m}(e_{2(i_0+m)} \right.$$
$$\left. - e_{2(i_0+m)+1}) + \sum_{i=m+i_0+1}^{\infty} a_i(e_{2i} - e_{2i+1}) \right\|$$
$$\geq \left\| \sum_{i \neq i_0} a_i(e_{2i} - e_{2i+1}) \right\| - \frac{2|a_{i_0}|}{m} \| e_1 \|$$

This gives the result when $m \to +\infty$.

We can now give the relations between $(x_n)_{n \in \mathbb{N}}$ and $(e_n)_{n \in \mathbb{N}}$:

Theorem II.3.4. Let $(x_n)_{n \in \mathbb{N}}$ be a good sequence in a separable Banach space X and $(e_n)_{n \in \mathbb{N}}$ the fundamental sequence of the associated spreading model. Denote by $\mathscr{X}$ this spreading model over X. If $(e_n)_{n \in \mathbb{N}}$ is not equivalent to the unit vector basis of ℓ_1, then

(i) $(e_n)_{n \in \mathbb{N}}$ is $\sigma(\mathscr{X}, \mathscr{X}^*)$-Cauchy.
(ii) $(x_n)_{n \in \mathbb{N}}$ is $\sigma(X, X^*)$-Cauchy.
(iii) $(e_n)_{n \in \mathbb{N}}$ is $\sigma(\mathscr{X}, \mathscr{X}^*)$-convergent if and only if $(x_n)_{n \in \mathbb{N}}$ is $\sigma(X, X^*)$-convergent and the limit is the same.

Proof

    (i) is an immediate consequence of Proposition II.3.1.

    (ii) Let $f \in X^*$ and suppose that there exist two subsequences $(x_{\varphi(n)})_{n \in \mathbb{N}}$ and $(x_{\psi(n)})_{n \in \mathbb{N}}$ of $(x_n)_{n \in \mathbb{N}}$ and two different real numbers λ and μ such that

$$\begin{cases} f(x_{\varphi(n)}) \xrightarrow[n \to +\infty]{} \lambda \\ f(x_{\psi(n)}) \xrightarrow[n \to +\infty]{} \mu \end{cases}$$

Extracting subsequences if necessary, we can suppose that for all $n \in \mathbb{N}$

$$\cdots < \varphi(n) < \psi(n) < \varphi(n + 1) < \psi(n + 1) < \cdots$$

Then, if we set $\tilde{f}(e_{2i}) = \lambda$, $\tilde{f}(e_{2i+1}) = \mu$, $\tilde{f}$ is a continuous linear form on the spreading model $\mathscr{X}$ over X which extends f. Since $(e_n)_{n \in \mathbb{N}}$ is $\sigma(\mathscr{X}, \mathscr{X}^*)$-Cauchy, this is impossible, and thus $(f(x_n))_{n \in \mathbb{N}}$ is convergent. This proves that $(x_n)_{n \in \mathbb{N}}$ is $\sigma(X, X^*)$-Cauchy.

    To prove (iii) we need two lemmas:

Lemma II.3.5. If $(x_n)_{n \in \mathbb{N}}$ is a good sequence in X and $x \in X$, then $(x_n - x)_{n \in \mathbb{N}}$ is a good sequence too, the two spreading models over X associated with $(x_n)_{n \in \mathbb{N}}$ and $(x_n - x)_{n \in \mathbb{N}}$ are isometric, and the two spreading models are isomorphic.

Proof of Lemma II.3.5. The first part is obvious.

    Denote by $(e_n)_{n \in \mathbb{N}}$ and $(f_n)_{n \in \mathbb{N}}$ the fundamental sequences of the spreading models defined by $(x_n)_{n \in \mathbb{N}}$ and $(x_n - x)_{n \in \mathbb{N}}$. The linear map T defined by

$$T\left(y + \sum_{i=0}^{k} a_i f_i \right) = y - \left(\sum_{i=0}^{k} a_i \right) x + \sum_{i=1}^{k} a_i e_i$$

is clearly an isometry from the spreading model over X defined by $(x_n - x)_{n \in \mathbb{N}}$ onto the spreading model over X defined by $(x_n)_{n \in \mathbb{N}}$.

    The last assertion is obvious too.

Lemma II.3.6. If $(x_n)_{n \in \mathbb{N}}$ is $\sigma(X, X^*)$-convergent to 0, then $(e_n)_{n \in \mathbb{N}}$ is basic, 1-unconditional.

Proof of Lemma II.3.6. We have to prove that, for all k, $i_0 \leq k$, and $a_{i_0} \neq 0$:

$$\left\| \sum_{\substack{i=0 \\ i \neq i_0}}^{k} a_i e_i \right\| \leq \left\| \sum_{i=0}^{k} a_i e_i \right\|$$

Let $\epsilon > 0$. There exists $N \in \mathbb{N}$ such that

$$n_0 > \cdots > n_k \geq N \Rightarrow \left| \left\| \sum_{\substack{i=0 \\ i \neq i_0}}^{k} a_i e_i \right\| - \left\| \sum_{\substack{i=0 \\ i \neq i_0}}^{k} a_i x_{n_i} \right\| \right| \leq \epsilon$$

$$\left| \left\| \sum_{i=0}^{k} a_i e_i \right\| - \left\| \sum_{i=0}^{k} a_i x_{n_i} \right\| \right| \leq \epsilon$$

Since 0 is in the $\sigma(X,X^*)$-closed and thus $\| \ \|$-closed convex hull of $(x_n)_{n \geq N}$ for all $N \in \mathbb{N}$, there exist $k_0 > \cdots > k_\ell \geq N$ and $p_0, \ldots, p_\ell$, and p in $\mathbb{N}$ with $\sum_{j=0}^{\ell} p_j = p$ such that

$$\left\| \frac{p_0 x_{k_0} + \cdots + p_\ell x_{k_\ell}}{p} \right\| \leq \frac{\epsilon}{|a_{i_0}|}$$

Fix a k-tuple $(n_j)_{j=0,\ldots,k, j \neq i_0}$ such that

$$n_k = N < n_{k-1} < \cdots < n_{i_0+1} < k_\ell < \cdots < k_0 < n_{i_0-1} < \cdots < n_0$$

Then we can write, for $k_\ell \leq j \leq k_0$,

$$\left\| \sum_{i=0}^{k} a_i e_i \right\| \geq \left\| \sum_{i=0}^{i_0-1} a_i x_{n_i} + a_{i_0} x_j + \sum_{i_0+1}^{k} a_i x_{n_i} \right\| - \epsilon$$

$$\geq \left\| \sum_{i=0}^{i_0-1} a_i x_{n_i} + a_{i_0} \frac{p_0 x_{k_0} + \cdots + p_\ell x_{k_\ell}}{p} + \sum_{i_0+1}^{k} a_i x_{n_i} \right\| - \epsilon$$

$$\geq \left\| \sum_{\substack{i=0 \\ i \neq i}}^{k} a_i x_{n_i} \right\| - 2\epsilon$$

$$\geq \left\| \sum_{\substack{i=0 \\ i \neq i}}^{k} a_i e_i \right\| - 3\epsilon$$

This gives the result when $\epsilon \to 0$.

Proof of (iii) of Theorem II.3.4. If $(x_n)_{n\in\mathbb{N}}$ is $\sigma(X,X^*)$-convergent to $x \in E$, then $(x_n - x)_{n\in\mathbb{N}}$ is $\sigma(X,X^*)$-convergent to 0 and $(e_n - x)_{n\in\mathbb{N}}$ is the fundamental sequence of the spreading model associated with $(x_n - x)_{n\in\mathbb{N}}$ by Lemma II.3.5. Thus, it is a 1-unconditional sequence by Lemma II.3.6. Since $(e_n)_{n\in\mathbb{N}}$ is not equivalent to the unit vector basis of ℓ_1, neither is $(e_n - x)_{n\in\mathbb{N}}$ and thus Corollary II.3.2. implies that $(e_n - x)$ is $\sigma(\mathscr{X},\mathscr{X}^*)$-convergent to 0.

On the other hand, suppose that $(e_n)_{n\in\mathbb{N}}$ is $\sigma(\mathscr{X},\mathscr{X}^*)$-convergent to e. Then there exist convex combinations $\theta_n = \sum_{i=p_n+1}^{p_{n+1}} \lambda_i e_i$ of $(e_n)_{n\in\mathbb{N}}$ that converge in norm to e. That is, for all $\epsilon > 0$, there exists $n_0 \in \mathbb{N}$ such that

$$\forall p,q \geq n_0, \quad \| \theta_p - \theta_q \| \leq \epsilon$$

But, by definition of the norm of the spreading model, there exists $x \in X$ such that

$$\forall p \geq n_0, \quad | \| \theta_p - \theta_{n_0} \| - \| \theta_p - x \| | \leq \epsilon$$

Then we get

$$\forall p \geq n_0, \quad \| \theta_p - x \| \leq 2\epsilon$$

So

$$e = x \in X$$

The sequence $(x_n - e)_{n\in\mathbb{N}}$ is $\sigma(\mathscr{X},\mathscr{X}^*)$-Cauchy by (ii). Moreover, let $f \in X^*$ and define $\tilde{f} \in \mathscr{X}^*$ by $\tilde{f}(e_i - e) = \lim_{n\to+\infty} f(x_n - e)$ for all $i \in \mathbb{N}$. Then $\tilde{f}$ extends f and since $\lim_{i\to+\infty} \tilde{f}(e_i - e) = 0$ we get $\lim_{n\to+\infty} f(x_n - e) = 0$ and this proves that $(x_n)_{n\in\mathbb{N}}$ is $\sigma(X,X^*)$-convergent to e.

The next result is an immediate consequence of Rosenthal's Theorem I.4.10 and Lemma II.3.5.

Corollary II.3.7. Let X be a separable Banach space.

(i) Every spreading model over X is isometric to a spreading model over X defined by a basic sequence.

(ii) Every spreading model is isomorphic to a spreading model defined by a basic sequence.

4. SPREADING MODELS ISOMETRIC TO ℓ_p $(1 \le p < \infty)$ OR c_0

In general, spreading models of a space X are not isomorphic to any subspace of this space. However, in the case where the spreading model is isometric to ℓ_p or c_0, then ℓ_p or c_0 is also isomorphic to a subspace of X. Let us give a precise definition and the result:

Definition II.4.1.   Let $(e_n)_{n \in \mathbb{N}}$ be the fundamental sequence of a spreading model over a separable Banach space X. We will say that the spreading model is *isometrically* ℓ_p $(1 \le p < \infty)$ *or* c_0 *over* X if for all finite sequences $(a_i)_{i \in \mathbb{N}}$ in $\mathbb{R}$ and all x in X, we have

$$\left\| x + \sum_{i=0}^{\infty} a_i e_i \right\| = \left\| x + \left(\sum_{i=0}^{\infty} |a_i|^p \right)^{1/p} e_1 \right\| \quad \text{if } 1 \le p < +\infty$$

$$= \| x + \sup_{i \in \mathbb{N}} |a_i| e_1 \| \qquad \text{if } p = +\infty$$

The following theorem is due to J. L. Krivine and B. Maurey [KM]. It is also given in [BeL]:

Theorem II.4.2.   Let $(x_n)_{n \in \mathbb{N}}$ be a good sequence in a separable Banach space X. Suppose that the spreading model over X, generated by $(e_n)_{n \in \mathbb{N}}$, is isometrically ℓ_p $(1 \le p < \infty)$ or c_0. Then for all $\epsilon > 0$ there exists a sequence $(x_{n_k})_{k \in \mathbb{N}}$ of $(x_n)_{n \in \mathbb{N}}$ that is $(1 + \epsilon)$-equivalent to the unit vector basis of ℓ_p $(1 \le p < \infty)$ or c_0.

Proof.   Suppose that the hypothesis of the theorem is true for some $p \in [1, +\infty[$; that is, $(e_n)_{n \in \mathbb{N}}$ is isometrically ℓ_p. The case of c_0 is analogous if we replace the sum $(\sum_{i=0}^{\infty} |a_i|^p)^{1/p}$ by $\text{Sup}_{i \in \mathbb{N}} |a_i|$. Fix $\epsilon > 0$ and a nonnegative sequence $(\epsilon_k)_{k \in \mathbb{N}}$ such that $\epsilon = \sum_{k=0}^{\infty} \epsilon_k$.

We proceed by induction. Suppose that $n_0 < n_1 < \cdots < n_k$ are already chosen such that for all finite sequences $(a_n)_{n \in \mathbb{N}}$ in $\mathbb{R}$, we have

$$\left| \left\| \sum_{i=0}^{k} a_i x_{n_i} + \left(\sum_{i=k+1}^{\infty} |a_i|^p \right)^{1/p} e_1 \right\| - \left(\sum_{i=0}^{\infty} |a_i|^p \right)^{1/p} \|e_1\| \right|$$

$$\le \left(\sum_{i=0}^{k} \epsilon_i \right) \left(\sum_{k=0}^{\infty} |a_i|^p \right)^{1/p}$$

Then we can choose n_{k+1} such that this property is true at the order

$k + 1$; indeed, let $(a^{(n)})_{n=0,\ldots,N}$ be an $(\epsilon_{k+1}/3\delta_{k+3}\, \mathrm{Sup}_{n\in\mathbb{N}}\,\|\,x_n\,\|)$-net in $S_{\ell_p^{k+3}}$, where δ_{k+3} is the isomorphism constant between ℓ_p^{k+3} and ℓ_1^{k+3}. For $n = 0,\ldots,N$, set $a^{(n)} = (a_i^{(n)})_{i=0,\ldots,k+2}$. By hypothesis we can find n_{k+1} such that, for all $n = 0,\ldots,N$,

$$\left|\ \left\|\sum_{i=0}^{k} a_i^{(n)} x_{n_i} + a_{k+1}^{(n)} x_{n_{k+1}} + a_{k+2}^{(n)} e_1\right\|\right.$$

$$\left.- \left\|\sum_{i=0}^{k} a_i^{(n)} x_{n_i} + (|\,a_{k+1}^{(n)}\,|^p + |\,a_{k+2}^{(n)}\,|^p)^{1/p}\, e_1\right\|\ \right| \le \frac{\epsilon_{k+1}}{3}$$

Then, if $a = (a_i)_{i\in\mathbb{N}}$ belongs to S_{ℓ_p}, there exists $n \in \{0, \ldots, N\}$ such that

$$\left|\ \left\|\sum_{i=0}^{k} a_i x_{n_i} + a_{k+1} x_{n_{k+1}} + \left(\sum_{i=k+2}^{\infty} |\,a_i\,|^p\right)^{1/p} e_1\right\|\right.$$

$$\left.- \left\|\sum_{i=0}^{k} a_i^{(n)} x_{n_i} + a_{k+1}^{(n)} x_{n_{k+1}} + a_{k+2}^{(n)} e_1\right\|\ \right|$$

$$\le \sum_{i=0}^{k+1} |\,a_i - a_i^{(n)}\,|\ \mathrm{Sup}_{n\in\mathbb{N}}\,\|\,x_n\,\|$$

$$+ \left|\left(\sum_{i=k+2}^{\infty} |\,a_i\,|^p\right)^{1/p} - a_{k+2}^{(n)}\right|\ \|\,e_1\,\|$$

$$\le \delta_{k+3}\left(\sum_{i=0}^{k+1} |\,a_i - a_i^{(n)}\,|^p\right.$$

$$+ \left|\left(\sum_{i=k+2}^{\infty} |\,a_i\,|^p\right)^{1/p} - a_{k+2}^{(n)}\right|^p\bigg)^{1/p} \mathrm{Sup}\,\|\,x_n\,\|$$

$$\le \frac{\epsilon_{k+1}}{3}$$

Since we can write

$$\left(\sum_{i=0}^{k} |\,a_i - a_i^{(n)}\,|^p + \left|\left(\sum_{i=k+1}^{\infty} |\,a_i\,|^p\right)^{1/p} - (|\,a_{k+1}^{(n)}\,|^p + |\,a_{k+2}^{(n)}\,|^p)^{1/p}\right|^p\right)^{1/p}$$

$$\le \left(\sum_{i=0}^{k+1} |\,a_i - a_i^{(n)}\,|^p\right.$$

$$+ \left|\left(\sum_{i=k+2}^{\infty} |\,a_i\,|^p\right)^{1/p} - a_{k+2}^{(n)}\right|^p\bigg)^{1/p}$$

for all n, we also have

$$\left| \; \left\| \sum_{i=0}^{k} a_i x_{n_i} + \left(\sum_{i=k+1}^{\infty} |a_i|^p \right)^{1/p} e_1 \right\| \right.$$
$$\left. - \left\| \sum_{i=0}^{k} a_i^{(n)} x_{n_i} + (|a_{k+1}^{(n)}|^p + |a_{k+2}^{(n)}|^p)^{1/p} e_1 \right\| \; \right| \leq \frac{\epsilon_{k+1}}{3}$$

Thus, for that n_{k+1}, we get by scaling, for all $(a_n)_{n\in\mathbb{N}}$ in ℓ_p,

$$\left| \; \left\| \sum_{i=0}^{k} a_i x_{n_i} + a_{k+1} x_{n_{k+1}} + \left(\sum_{i=k+2}^{\infty} |a_i|^p \right)^{1/p} e_1 \right\| \right.$$
$$\left. - \left\| \sum_{i=0}^{k} a_i x_{n_i} + \left(\sum_{i=k+1}^{\infty} |a_i|^p \right)^{1/p} e_1 \right\| \; \right| \leq \epsilon_{k+1} \left(\sum_{i=0}^{\infty} |a_i|^p \right)^{1/p}$$

By the hypothesis, we know that

$$\left| \; \left\| \sum_{i=0}^{k} a_i x_{n_i} + \left(\sum_{i=k+1}^{\infty} |a_i|^p \right)^{1/p} e_1 \right\| - \left(\sum_{i=0}^{\infty} |a_i|^p \right)^{1/p} \|e_1\| \; \right|$$
$$\leq \left(\sum_{i=0}^{k} \epsilon_i \right) \left(\sum_{i=0}^{\infty} |a_i|^p \right)^{1/p}$$

Thus we get

$$\left| \; \left\| \sum_{i=0}^{k+1} a_i x_{n_i} + \left(\sum_{i=k+2}^{\infty} |a_i|^p \right)^{1/p} e_1 \right\| - \left(\sum_{i=0}^{\infty} |a_i|^p \right)^{1/p} \|e_1\| \; \right|$$
$$\leq \left(\sum_{i=0}^{k+1} \epsilon_i \right) \left(\sum_{i=0}^{\infty} |a_i|^p \right)^{1/p}$$

This proves that our induction is true and finishes the proof.

5. KRIVINE'S THEOREM

This famous theorem on the finite representability of c_0 or ℓ_p in Banach spaces was proved by J. L. Krivine in [K2]. The proof was simplified and the result slightly improved by H. P. Rosenthal in [Ros7]. We present here a simpler proof due to H. Lemberg [Le] which takes up some of the ideas of [Ros7]. See also [BeL] for the complex version of this theorem.

Theorem II.5.1. Let X be a Banach space and $(x_k)_{k \in \mathbb{N}}$ a good sequence in X. Then there exists $p \in [1, +\infty]$ such that ℓ^p (or c_0 if $p = +\infty$) is block f.r. in $(x_n)_{n \in \mathbb{N}}$.

The proof of this theorem is long and rather difficult. It will need several definitions and lemmas. In particular, we will have to use some operator theory (see [DS] for background) and this obliges us to complexify X:

Definition II.5.2. A sequence $(y_n)_{n \in \mathbb{N}}$ is said to be *K-sign-unconditional* if for every $(\epsilon_n)_{n \in \mathbb{N}}$ with $\epsilon_n = \pm 1$ for all n, $(\epsilon_n y_n)_{n \in \mathbb{N}}$ is K-equivalent to $(y_n)_{n \in \mathbb{N}}$.

Remark. If $(y_n)_{n \in \mathbb{N}}$ is K-sign-unconditional, it is K-unconditional, and if $(y_n)_{n \in \mathbb{N}}$ is K-unconditional, it is $2K$-sign-unconditional (see Theorem I.3.2. and Definition I.3.4).

Definition II.5.3. If X is a Banach space with a 1-sign-unconditional basis $(e_n)_{n \in \mathbb{N}}$, we will define its *complexification* $X^{\mathbb{C}}$ as the closed complex linear span of $(e_n)_{n \in \mathbb{N}}$ under the norm:

$$\forall (a_n)_{n \in \mathbb{N}} \in \mathbb{C}^{(\mathbb{N})}, \qquad \left\| \sum_{i=0}^{\infty} a_i e_i \right\| + \left\| \sum_{i=0}^{\infty} | a_i | e_i \right\|$$

It is easy to see that this defines a norm on $X^{\mathbb{C}}$ and that it induces the original norm on X.

For $Z = \sum_{i=0}^{\infty} a_i e_i \in X^{\mathbb{C}}$, we will define its *modulus* by $| Z | = \sum_{i=0}^{\infty} | a_i | e_i$, its *real part* by $\mathrm{Re}\, Z = \sum_{i=0}^{\infty} (\mathrm{Re}\, a_i) e_i$, and so on.

Let us set two lemmas on operator theory:

Lemma II.5.4. Let $X^{\mathbb{C}}$ be a complex Banach space and T a linear operator on $X^{\mathbb{C}}$. If λ is in the boundary of the spectrum of T, then there exists a normalized sequence $(u_n)_{n \in \mathbb{N}}$ in $X^{\mathbb{C}}$ such that $Tu_n - \lambda u_n \to 0$ when $n \to +\infty$.

Proof of Lemma II.5.4. Changing T to $T - \lambda I$, if necessary, we can suppose that $\lambda = 0$.

Suppose that the conclusion of the lemma is false. Then there exists

$C > 0$ such that for all $x \in X$, $\| Tx \| \geq C \| x \|$. But, since 0 is in the boundary of the spectrum of T, there exists μ, $| \mu | < C/2$ such that $T - \mu I$ is invertible. Set

$$Q = \frac{1}{\mu} \sum_{n=0}^{\infty} (-1)^n \mu^n (T - \mu I)^{-n}$$

This series is norm convergent because we can write

$$\| (-1)^n \mu^n (T - \mu I)^{-n} \| \leq | \mu |^n \| (T - \mu I) \|^{-n} \leq | \mu |^n \left(\frac{2}{C} \right)^n$$

Thus Q is continuous on $X^{\mathbb{C}}$ and moreover we have $QT = TQ = I$. This contradicts the fact that 0 is in the spectrum of T. (We recall that the spectrum of an operator is a closed set in $\mathbb{C}$.) This proves Lemma II.5.4.

Lemma II.5.5. Let $X^{\mathbb{C}}$ be a complex Banach space and T, S belong to $\mathcal{L}(X)$ such that $TS = ST$. If λ is in the boundary of the spectrum of T, there exist μ in the spectrum of S and a normalized sequence $(w_n)_{n \in \mathbb{N}}$ in $X^{\mathbb{C}}$ such that

$$\begin{cases} Tw_n - \lambda w_n \to 0 & \text{when } n \to +\infty \\ Sw_n - \mu w_n \to 0 & \text{when } n \to +\infty \end{cases}$$

Proof of Lemma II.5.5. Let $(u_n)_{n \in \mathbb{N}}$ be given by Lemma II.5.4, applied to T and λ. If $\tilde{X}^{\mathbb{C}} = (X^{\mathbb{C}})/\mathcal{U}$ is an ultrapower of $X^{\mathbb{C}}$, we extend T and S to $\tilde{X}^{\mathbb{C}}$ by

$$\forall \tilde{x} = (x_n)_{n \in \mathbb{N}} \in \tilde{X}^{\mathbb{C}}, \qquad \tilde{T}\tilde{x} = (Tx_n)_{n \in \mathbb{N}}$$

$$\tilde{S}\tilde{x} = (Sx_n)_{n \in \mathbb{N}}$$

It is clear that $\tilde{T}$ and $\tilde{S}$ are continuous on $\tilde{X}$ and commute. Set $\tilde{Z} = \{\tilde{z} \in \tilde{X}^{\mathbb{C}} / \tilde{T}\tilde{z} = \lambda \tilde{z}\}$. Then $\tilde{Z}$ is a subspace of $\tilde{X}^{\mathbb{C}}$, not equal to $\{0\}$ because $\tilde{u} = (u_n)_{n \in \mathbb{N}}$ belongs to $\tilde{Z}$ and is of norm 1. Since $\tilde{T}$ and $\tilde{S}$ commute, $\tilde{S}\tilde{Z} \subset \tilde{Z}$ and we can consider the operator $\tilde{S}|_{\tilde{Z}}$ on $\tilde{Z}$. Let μ belong to the boundary of the spectrum of $\tilde{S}/\tilde{z}$. Note that μ belongs to the spectrum of S.

If $(\tilde{u}'_n)_{n \in \mathbb{N}}$ is given by Lemma II.5.4, applied to $\tilde{S}|_{\tilde{Z}}$ and μ, we have

$$\begin{cases} \tilde{T}\tilde{u}'_n = \lambda \tilde{u}'_n & \text{for all } n \in \mathbb{N} \\ \tilde{S}\tilde{u}'_n - \mu \tilde{u}'_n \to 0 & \text{when } n \to +\infty \\ \| \tilde{u}'_n \| = 1 & \text{for all } n \in \mathbb{N} \end{cases}$$

Thus, if we define $\tilde{X}^{\mathbb{C}} = \bar{X}^{\mathbb{C}}/\mathfrak{U}$ and $\tilde{T}, \tilde{S}$ as above, we have proved the existence of $\bar{u}' = (\bar{u}'_n)_{n \in \mathbb{N}} \in \tilde{X}^{\mathbb{C}}$ such that

$$\begin{cases} \tilde{\tilde{T}}\tilde{\tilde{u}}' = \lambda \tilde{\tilde{u}}' \\ \tilde{\tilde{S}}\tilde{\tilde{u}}' = \mu \tilde{\tilde{u}}' \\ \| \tilde{\tilde{u}}' \| = 1 \end{cases}$$

If we take representatives of $\tilde{\tilde{u}}'$ in X, that is, $\tilde{\tilde{u}}' = (w_n)_{n \in \mathbb{N}}$ with $\| w_n \| = 1$ for all n in $\mathbb{N}$, it is not hard to see that there exist $(n_k)_{k \in \mathbb{N}}$, increasing sequence of integers such that $(w_{n_k})_{k \in \mathbb{N}}$ verifies the conclusion of Lemma II.5.5.

Before proving Theorem II.5.1, we are going to reduce the study to a simpler case:

Let $(x_n)_{n \in \mathbb{N}}$ be a good sequence in X. We may assume X to be separable; indeed, if not, we replace X by $\overline{\text{Span}} [x_n, n \in \mathbb{N}]$.

Let $(e_n)_{n \in \mathbb{N}}$ be the fundamental sequence of the spreading model, defined by $(x_n)_{n \in \mathbb{N}}$. Since $(e_n)_{n \in \mathbb{N}}$ is block f.r. in $(x_n)_{n \in \mathbb{N}}$, (see Definition II.2.4) it is sufficient to prove this theorem for $(e_n)_{n \in \mathbb{N}}$ instead of $(x_n)_{n \in \mathbb{N}}$.

The same argument shows that we can replace $(e_n)_{n \in \mathbb{N}}$ by $(e_{2n+1} - e_{2n})_{n \in \mathbb{N}}$. Thus, by Proposition II.3.3, we can suppose without loss of generality that $(e_n)_{n \in \mathbb{N}}$ is 1-unconditional.

We need a little more, namely the 1-sign-unconditionality:

Lemma II.5.6. Let $(e_n)_{n \in \mathbb{N}}$ be an I.S. and 1-unconditional sequence and $\mathfrak{U}$ a nontrivial ultrafilter on $\mathbb{N}$.

1. If for all $n \in \mathbb{N}$, $(b_j^n)_{j \in \mathbb{N}}$ is a sequence of normalized blocks on $(e_n)_{n \in \mathbb{N}}$, we define: for $(a_n)_{n \in \mathbb{N}} \in \mathbb{R}^{(\mathbb{N})}$, $\| \sum_{i=0}^{\infty} a_i v_i \| = \lim_{n,\mathfrak{U}} \| \sum_{i=0}^{\infty} a_i b_i^n \|$. Then $\| \sum_{i=0}^{\infty} a_i v_i \|$ is a norm on $\mathbb{R}^{(\mathbb{N})}$, $(v_i)_{i \in \mathbb{N}}$ is I.S., 1-unconditional, and block f.r. in $(e_n)_{n \in \mathbb{N}}$.

2. For all $n \in \mathbb{N}$, there exist normalized blocks $(b_i^n)_{i \in \mathbb{N}}$ on $(e_n)_{n \in \mathbb{N}}$ such that $(v_i)_{i \in \mathbb{N}}$ is 1-sign-unconditional.

Proof of Lemma II.5.6. The first part is immediate.

To prove 2, we have to distinguish between two cases, depending on the value of $\text{Sup}_n \| \sum_{i=0}^{n} e_i \|$:

First case: $\text{Sup}_n \| \sum_{i=0}^{n} e_i \| < +\infty$. Then $(e_i)_{i \in \mathbb{N}}$ is equivalent to the unit vector basis of c_0 because of the 1-unconditionality (see Theorem

I.3.2). Then, for $n \in \mathbb{N}^*$, we can apply James's theorem (see Theorem I.4.7) with $\epsilon = 1/n$; we obtain a sequence that is $(1 + 1/n)$-equivalent to the unit vector basis of c_0. By a standard argument on blocks (see Corollary I.1.18), we can suppose that there exist blocks $(b_j^n)_{j \in \mathbb{N}}$ on $(e_n)_{n \in \mathbb{N}}$ that are also $(1 + 1/n)$-equivalent to the unit vector basis of c_0. Thus, by definition, $(v_i)_{i \in \mathbb{N}}$ is 1-equivalent to the unit vector basis of c_0; therefore it is 1-sign-unconditional.

Second case: $\mathrm{Sup}_n \| \sum_{i=0}^n e_i \| = +\infty$. For n and $i \in \mathbb{N}$, set $b_i^n = \sum_{j=0}^{n-1} (-1)^j e_{j+ni} / \| \sum_{j=0}^{n-1} (-1)^j e_j \|$. By the remark following Definition II.5.2, we know that

$$\left\| \sum_{j=0}^{n-1} e_j \right\| \le 2 \left\| \sum_{j=0}^{n-1} (-1)^j e_j \right\|$$

If $\epsilon_i = \pm 1$ for all $i \in \mathbb{N}$, we want to show that, for all fixed $(a_0, \ldots, a_m) \in \mathbb{R}^{m+1}$,

$$\lim_{n, \mathcal{U}} \left| \left\| \sum_{i=0}^m a_i b_i^n \right\| - \left\| \sum_{i=0}^m \epsilon_i a_i b_i^n \right\| \right| = 0$$

For $n, i \in \mathbb{N}$, define $b_i'^n = \sum_{j=0}^{n-1} (-1)^j e_{j+ni+1} / \| \sum_{j=0}^{n-1} (-1)^j e_j \|$. Observe that $\| b_i^n - b_i'^n \| \le 2 \| e_1 \| / \| \sum_{j=0}^{n-1} e_j \|$. Then, since $(e_n)_{n \in \mathbb{N}}$ is I.S., we can write

$$\left\| \sum_{i=0}^m a_i b_i^n \right\| = \left\| \sum_{i=0}^m a_i b_{2i}^n \right\| = \left\| \sum_{i=0}^m a_i b_{2i}'^n \right\|$$

$$\left\| \sum_{i=0}^m \epsilon_i a_i b_i^n \right\| = \left\| \sum_{\epsilon_i > 0} a_i b_{2i}^n - \sum_{\epsilon_i < 0} a_i b_{2i}'^n \right\|$$

Then

$$\left| \left\| \sum_{i=0}^m a_i b_i^n \right\| - \left\| \sum_{i=0}^m \epsilon_i a_i b_i^n \right\| \right| \le \left\| \sum_{\epsilon_i < 0} a_i (b_{2i}^n - b_{2i}'^n) \right\|$$

$$\le \mathrm{Sup}_{i \in \mathbb{N}} | a_i | \frac{4 \| e_1 \|}{\left\| \sum_{j=0}^{n-1} e_j \right\|}$$

This tends to 0 when n tends to ∞ and the proof is complete.

Proof of Theorem II.5.1. Changing $(x_n)_{n \in \mathbb{N}}$ in $(v_n)_{n \in \mathbb{N}}$ given by Lemma II.5.6 if necessary, we can and will suppose that $(x_n)_{n \in \mathbb{N}}$ is I.S., 1-unconditional and 1-sign-unconditional.

In the sequel, we are going to change the indexation of the sequence $(x_n)_{n \in \mathbb{N}}$: let us change $\mathbb{N}$ into $\mathbb{Q}^+$ and consider the sequence $(x_d)_{d \in \mathbb{Q}^+}$. Precisely, if $0 \leq d_0 \leq \cdots \leq d_k$ belong to $\mathbb{Q}^+$, then for all $(a_0, \ldots, a_k)$ in $\mathbb{R}^{k+1}$ we define

$$\| a_0 x_{d_0} + \cdots + a_k x_{d_k} \| = \| a_0 x_0 + \cdots + a_k x_k \|$$

Call $X = \overline{\text{Span}} [x_d, d \in \mathbb{Q}^+]$ and for all m in $\mathbb{N}$, denote by $X_{[m,m+1[}$ the space $\overline{\text{Span}} [x_d, d \in \mathbb{Q}^+ \cap [m, m+1[$. For all $m \in \mathbb{N}$, we define two operators T and S on $X_{[m,m+1[}$ as follows:

For all finite sequences $(a_i)_{i \in \mathbb{N}}$ in $\mathbb{R}$ and all sequences $(d_i)_{i \in \mathbb{N}}$ in $[m, m+1[\cap \mathbb{Q}^+$,

$$T \left(\sum_{i=0}^{\infty} a_i x_{d_i} \right) = \sum_{i=0}^{\infty} a_i x_{(d_i + m)/2} + \sum_{i=0}^{\infty} a_i x_{(d_i + m + 1)/2}$$

$$S \left(\sum_{i=0}^{\infty} a_i x_{d_i} \right) = \sum_{i=0}^{\infty} a_i x_{(d_i + 2m)/3} + \sum_{i=0}^{\infty} a_i x_{(d_i + 2m + 1)/3}$$

$$+ \sum_{i=0}^{\infty} a_i x_{(d_i + 2m + 2)/3}$$

Then T and S also define operators on X and they commute. These operators can be extended to $X^{\mathbb{C}}$ and $X_{[m,m+1[}^{\mathbb{C}}$ for all m, in a natural way by the same formulas with coefficients in $\mathbb{C}$.

We need two lemmas:

Lemma II.5.7. There exist two positive numbers λ and μ and a normalized sequence $(f^n)_{n \in \mathbb{N}}$ in $X_{[0,1[}$ such that

$$\| Tf^n - \lambda f^n \| \to 0 \quad \text{when } n \to +\infty$$
$$\| Sf^n - \mu f^n \| \to 0 \quad \text{when } n \to +\infty$$

Proof of Lemma II.5.7. We apply Lemma II.5.5 with T and S as operators on $X_{[0,1[}^{\mathbb{C}}$. We get two complex numbers λ and μ and a normalized sequence $(f^n)_{n \in \mathbb{N}}$ in $X_{[0,1]}^{\mathbb{C}}$ that verify

$$\| Tf^n - \lambda f^n \| \to 0 \quad \text{when } n \to +\infty$$
$$\| Sf^n - \mu f^n \| \to 0 \quad \text{when } n \to +\infty$$

By definition of T and S and of the complexification of $X_{[0,1[}$, we have

$$\| T | f^n | - | \lambda | | f^n | \| \le \| Tf^n - \lambda f^n \|$$
$$\| S | f^n | - | \mu | | f^n | \| \le \| Sf^n - \mu f^n \|$$

(These inequalities give true coordinates as coordinates on the basis $(x_d)_{d \in \mathbb{Q}^+}$ by the triangle inequality in $\mathbb{C}$; then apply Definition II.5.3.)

Thus $| \lambda |$, $| \mu |$, and $(| f^n |)_{n \in \mathbb{N}}$ verify the conclusion of Lemma II.5.7.

Lemma II.5.8.   For all $f \in X$,

$$\| f \| \le \| Tf \| \le 2 \| f \|$$
$$\| f \| \le \| Sf \| \le 3 \| f \|$$

Proof of Lemma II.5.8.   It is an obvious consequence of the fact that $(x_d)_{d \in \mathbb{Q}^+}$ is I.S. and 1-sign-unconditional by definition.

We come back to the proof of Theorem II.5.1. By Lemma II.5.7, there exist λ and $\mu \ge 0$ and a normalized sequence $(f^n)_{n \in \mathbb{N}}$ in $X_{[0,1[}$ such that

$$\| Tf^n - \lambda f^n \| \to 0 \quad \text{when } n \to +\infty$$
$$\| Sf^n - \mu f^n \| \to 0 \quad \text{when } n \to +\infty$$

Lemma II.5.8 implies that

$$1 \le \lambda \le 2$$
$$1 \le \mu \le 3$$

By an easy approximation of f^n by an element of $X_{[0,1[}$ with finite nonzero coordinates on $(x_d)_{d \in \mathbb{N}}$, we can suppose that for all $n \in \mathbb{N}$, f^n has a decomposition of the form

$$f^n = \sum_{i=0}^{k_n} a_i^n x_{d_i^n} \quad \text{with } a_i^n \in \mathbb{R} \text{ and } d_i^n \in [0,1[\cap \mathbb{Q} \text{ for all } i \in \mathbb{N}$$

Define, for all $n \in \mathbb{N}$,

$$g_m^n = \sum_{i=0}^{k_n} a_i^n x_{d_i^n + m} \in X_{[m,m+1[}$$

Let $\mathcal{U}$ be a nontrivial ultrafilter on $\mathbb{N}$.

We define a new norm on $\mathbb{R}^{(\mathbb{N})}$ by

$$\forall (a_i)_{\in \mathbb{N}} \subset \mathbb{R}^{(\mathbb{N})}, \qquad \left\| \sum_{i=0}^{\infty} a_i u_i \right\| = \lim_{n, \mathcal{U}} \left\| \sum_{i=0}^{\infty} a_i g_i^n \right\|$$

Observe that, since $(x_d)_{d \in \mathbb{Q}^+}$ is I.S., 1-unconditional, and 1-sign-unconditional, $(u_m)_{m \in \mathbb{N}}$ also has these properties. Moreover, since $(g_i^n)_{i \in \mathbb{N}}$ are blocks on $(x_d)_{d \in \mathbb{Q}^+}$, $(u_m)_{m \in \mathbb{N}}$ is block f.r. in $(x_d)_{d \in \mathbb{Q}^+}$.

We are going to prove that $(u_m)_{m \in \mathbb{N}}$ is 1-equivalent to the unit vector basis of c_0 or ℓ_p for some $p \in [1, +\infty[$. For this, we will need one more lemma:

Lemma II.5.9. Let $(u,v) \in \overline{\text{Span}} \, [u_m, \, m \in \mathbb{N}]$ such that

$$\text{Supp } u < j < j + 1 < j + 2 < \text{Supp } v$$

Then

$$\| u + \lambda u_j + v \| = \| u + u_j + u_{j+1} + v \|$$

$$\| u + \mu u_j + v \| = \| u + u_j + u_{j+1} + u_{j+2} + v \|$$

(Supp u has been defined in Definition I.1.1.)

Proof of Lemma II.5.9. By definition of T and S and since $(e_d)_{d \in \mathbb{Q}^+}$ is I.S., it is clear that, for all $(f,g) \in X^{\mathbb{C}}$ such that

$$\text{Supp } f < \text{Supp } Tg_m^n < \text{Supp } g$$

$$\text{Supp } f < \text{Supp } Sg_m^n < \text{Supp } g$$

$$\text{Supp } f < \text{Supp } (g_m^n + g_{m+1}^n + g_{m+2}^n) < \text{Supp } g$$

then

$$\| f + Tg_m^n + g \| = \| f + g_m^n + g_{m+1}^n + g \|$$

$$\| f + Sg_m^n + g \| = \| f + g_m^n + g_{m+1}^n + g_{m+2}^n + g \|$$

In these inequalities, if we take limits with respect to $\mathcal{U}$, we get the result.

By an easy induction, we obtain the following corollary:

Corollary II.5.10. If Supp $u < j < j + 2^k 3^l <$ Supp v, then

$$\| u + \lambda^k \mu^l u_j + v \| = \| u + u_j + u_{j+1} + \cdots u_{j + 2^k 3^l - 1} + v \|$$

The following corollary will finish the proof of Theorem II.5.1:

Corollary II.5.11. Either $(u_m)_{m \in \mathbb{N}}$ is 1-equivalent to the unit vector basis of c_0 or there exists $p \in [1, +\infty[$ such that $(u_m)_{m \in \mathbb{N}}$ is 1-equivalent to the unit vector basis of ℓ_p.

Proof of Corollary II.5.11. Since $(u_m)_{m \in \mathbb{N}}$ is 1-unconditional, Corollary II.5.10 implies

$$2^k < 3^l \Rightarrow \lambda^k \leq \mu^l$$
$$2^k > 3^l \Rightarrow \lambda^k \geq \mu^l$$

In particular, we have

$$\lambda = 1 \Rightarrow \mu = 1 \quad \text{and} \quad \lambda > 1 \Rightarrow \mu > 1$$

We have to distinguish between two cases, depending on the value of λ:

First case: $\lambda = 1$. Then $\mu = 1$ too and then by Corollary II.5.11. we get, for all $N \in \mathbb{N}$,

$$\left\| \sum_{i=0}^{N} u_i \right\| = 1$$

So, for all $(a_i)_{i \leq N} \in \mathbb{R}^N$,

$$\left\| \sum_{i=0}^{N} a_i u_i \right\| \geq \text{Sup} \{| a_i |, i \leq N\}$$

and

$$\leq \left\| \sum_{i=0}^{N} \text{Sup} \{| a_i |, i \leq N\} u_i \right\|$$

$$\leq \text{Sup} \{| a_i |, i \leq N\} \left\| \sum_{i=0}^{N} u_i \right\|$$

$$\leq \text{Sup} \{| a_i |, i \leq N\}$$

Thus, $(u_m)_{m \in \mathbb{N}}$ is 1-equivalent to the unit vector basis of c_0

Second case: $\lambda > 1$. Then $\mu > 1$ too and since $\lambda \leq 2$, $\mu \leq 3$, there exist p, q in $[1, +\infty[$ such that

$$\lambda = 2^{1/p} \quad \text{and} \quad \mu = 3^{1/q}$$

Suppose that $p \neq q$; for example, take $p > q$. Choose r and s in $\mathbb{N}$ such that

$$3^r < 2^s < 3^{rp/q}$$

Then we get both following properties together: $3^r < 2^s$ and $\mu^r > \lambda^s$, which is a contradiction.

Thus $p = q$ and for k, l in $\mathbb{N}$ and by Corollary II.5.10, we get

$$\left\| \sum_{i=0}^{2^k 3^l - 1} u_i \right\| = (2^k 3^l)^{1/p}$$

Let us admit for a while the following standard number-theoretic result, which is easily understandable and which will be proved at the end of the proof of Theorem II.5.1.

Lemma II.5.12. For all $N \in \mathbb{N}$ and $\epsilon > 0$, there exist k, l, k', l' in $\mathbb{N}$ such that

$$2^k 3^{-l} \leq N \leq 2^{k'} 3^{-l'} \leq (1 + \epsilon) 2^k 3^{-l}$$

Then, for all $N \in \mathbb{N}$ we can call $L = \text{Sup}(l, l')$; we get

$$\left\| \sum_{i=0}^{2^k 3^{L-l}} u_i \right\| = (2^k 3^{L-l})^{1/p} \leq \left\| \sum_{i=0}^{3^L N - 1} u_i \right\| \leq \left\| \sum_{i=0}^{2^{k'} 3^{L-l'} - 1} u_i \right\|$$

$$\leq (2^{k'} 3^{L-l'})^{1/p} \leq (1 + \epsilon)^{1/p} (2^k 3^{L-l})^{1/p}$$

When ϵ tends to 0, this proves that

$$\left\| \sum_{i=0}^{3^L N - l} u_i \right\| = (3^L)^{1/p} \left\| \sum_{i=0}^{N-1} u_i \right\| = (3^L)^{1/p} N^{1/p}$$

And thus

$$\left\| \sum_{i=0}^{N-1} u_i \right\| = N^{1/p}$$

Now, if we take for $i = 0, \ldots, N - 1$,

$$a_i = 2^{k_i} 3^{-l_i}, \qquad L = \text{Sup}\{l_i, i = 0, \ldots, N - 1\}, \qquad M = \sum_{i=0}^{N-1} 2^{k_i} 3^{L-l_i}$$

we get

$$
\left\| \sum_{i=0}^{N-1} a_i^{1/p} 3^{L/p} u_i \right\| = \left\| \sum_{i=0}^{N-1} 2^{k_i/p} 3^{L - l_i/p} u_i \right\|
$$

$$
= \left\| \sum_{i=0}^{M-1} u_i \right\| = M^{1/p}
$$

$$
= \left(\sum_{i=0}^{N-1} a_i 3^L \right)^{1/p}
$$

Thus:

$$
\left\| \sum_{i=0}^{N-1} a_i^{1/p} u_i \right\| = \left(\sum_{i=0}^{N-1} a_i \right)^{1/p}
$$

We can conclude that $(u_m)_{m \in \mathbb{N}}$ is 1-equivalent to the unit vector basis of ℓ_p by 1-sign-unconditionality and density.

This proves Corollary II.5.11 and finishes the proof of Theorem II.5.1.

Proof of Lemma II.5.12. We use Rosenthal's proof [Ros7]: By taking logarithms, it is sufficient to show that the set $H = \{k\tau - l, k \in \mathbb{N}, l \in \mathbb{N}\}$ where $\tau = (\log 2)/(\log 3)$, is dense in $\mathbb{R}$. Since τ is irrational, there is an infinite sequence $(p_i/q_i)_{i \in \mathbb{N}}$ in $\mathbb{Q}$ such that $|\tau - p_i/q_i| \leq 1/q_i$ for all $i \in \mathbb{N}$. Thus $q_i\tau - p_i$ belongs to $H \cap [0,1]$ for all $i \in \mathbb{N}$ and $H \cap [0,1]$ is infinite.

By compactness, there exists an infinite Cauchy sequence $(h_i)_{i \in \mathbb{N}}$ in the set $H \cap [0,1]$. Note that $h_i = k_i\tau - l_i$ for all $i \in \mathbb{N}$. Since the sequence $(h_i)_{i \in \mathbb{N}}$ is infinite, the two sequences $(k_i)_{i \in \mathbb{N}}$ and $(l_i)_{i \in \mathbb{N}}$ defined above are not bounded. Thus, for all $\epsilon > 0$, there exist i and j in $\mathbb{N}$ such that $0 < |h_i - h_j| \leq \epsilon$ and $h_i - h_j = (k_i - k_j)\tau + (l_i - l_j)$ belongs to H. Define $h^1 = h_i - h_j$ and suppose for instance that h^1 is positive. Then there exists $n \in \mathbb{N}$ such that $-\epsilon \leq nh^1 - 1 < 0$. Define $h^2 = nh_1 - 1$ and remark that h^2 belongs to H. We have proved that, for all $\epsilon > 0$, there exist h^1 and h^2 in H with $0 < h^1 \leq \epsilon$ and $-\epsilon \leq h^2 < 0$.

Since for all $n \in \mathbb{N}$ and $h \in H$, nh belongs to H, this proves that H is dense in $\mathbb{R}^+$ and $\mathbb{R}^-$ and thus in $\mathbb{R}$.

A consequence of Krivine's theorem is the following:

Theorem II.5.13. Let X be a Banach space. Suppose that there exist $p \in [1, +\infty]$ and $K > 0$ such that for all $n \in \mathbb{N}$, there is a subspace of X which is K-isomorphic to ℓ_p^n. Then ℓ_p is f.r. in X (c_0 is f.r. in X if $p = +\infty$).

Proof. It is sufficient to show that ℓ_p (resp. c_0) is f.r. in any space Y that is K-isomorphic to ℓ_p (resp. c_0). Let us give the proof in the case of ℓ_p:

Let Y be such a space and set $Y = \text{Span} [x_n, n \in \mathbb{N}]$. Theorem II.5.1 shows that there exists $q \in [1, +\infty]$ such that for all $n \in \mathbb{N}$ and $\epsilon > 0$, there exist blocks $(u_i)_{i \in \mathbb{N}}$ on $(x_i)_{i \in \mathbb{N}}$, that are $(1 + \epsilon)$-equivalent to ℓ_q^n. Since by hypothesis these blocks are K-equivalent to ℓ_p^n, we get $p = q$ and ℓ_p is f.r. in Y.

Remarks II.5.14. If one reads the proof of Theorem II.5.1 carefully, one can improve the statement in two directions:

1. Let us call $(u_i^{(n,\epsilon)})_{0 \leq i \leq n-1}$ n blocks on $(x_k)_{k \in \mathbb{N}}$ that are $(1 + \epsilon)$-equivalent to ℓ_p^n according to Theorem II.5.1; then, it can be shown that these blocks have *the same coefficients* on $(x_k)_{k \in \mathbb{N}}$. That is, there exist $N \in \mathbb{N}$ and scalars $(a_i^{(n,\epsilon)})_{0 \leq i \leq N}$ such that:

$$\text{for all } j = 0, \ldots, n-1, \quad u_j^{(n,\epsilon)} = \sum_{i=0}^{N} a_i^{(n,\epsilon)} x_{i+j(N+1)}$$

2. Corollary II.5.14 is true over the space; that is, for all x in X and all finite sequence $(a_i)_{i \in \mathbb{N}}$ in $\mathbb{R}$,

$$\left\| x + \sum_{i=0}^{\infty} a_i u_i \right\| = \left\| x + \left(\sum_{i=0}^{\infty} |a_i|^p \right)^{1/p} u_i \right\| \qquad \text{if } p < +\infty$$
$$= \| x + \text{Sup} \{ |a_i|, i \in \mathbb{N}\} u_i \| \quad \text{if } p = +\infty$$

NOTES AND REMARKS

1. An important superproperty is superreflexivity. Let us mention a deep result of P. Enflo [E2], which was improved later by G. Pisier [P5]: a Banach space X is superreflexive if and only if there is an equivalent norm on X that is uniformly convex. This result, as well as other characterizations of superreflexivity in terms of martin-

gales with values in X or in terms of finite trees in B_x, due to R. C. James [J5, J6, J8], can be found in [Be3].

2. Some other properties of spreading models are known, for example, the Banach-Saks property, which was studied first by A. Brunel and L. Sucheston [BS1, BS2] and then by H. P. Rosenthal and B. Beauzamy; it is closely related to the behavior of fundamental sequences of certain spreading models of X. See [Be2], [Be3], [Be4], [BeL], [BM], [GL1], [ROS9], for more details and further applications.

3. There is a complex version of Krivine's theorem that is given in [BeL].

4. In Chapter IV it will be proved that ℓ_2 is finitely representable in ℓ_p, $1 \leq p \leq +\infty$, and c_0. As a consequence of this result and Krivine's theorem (II.5.1), we immediately get the following result, due to A. Dvoretsky [Dv]: ℓ_2 is f.r. in any Banach space X. This theorem has a lot of applications in the local theory of Banach spaces.

EXERCISES

Most of the results that are given in these exercises can be found in [BeL].

1. Let X be a Banach space with a basis $(e_n)_{n \in \mathbb{N}}$ and $(x_n)_{n \in \mathbb{N}}$ a good sequence in X, which converges to 0 for $\sigma(X, X^*)$. Prove that there exists a sequence of blocks on $(e_n)_{n \in \mathbb{N}}$ that defines the same spreading model as $(x_n)_{n \in \mathbb{N}}$.

2. Prove that X has a spreading model isomorphic to ℓ_1 if and only if it has a spreading model whose fundamental sequence is equivalent to the unit vector basis of ℓ_1.

3. Let $1 \leq p < +\infty$. Show that every spreading model over ℓ_p is isometric to ℓ_p.

4. (a) Show that c_0 has exactly two I.S. basic sequences: the unit vector basis and the summing basis $(s_n)_{n \in \mathbb{N}}$ defined by $s_n = \sum_{i=0}^{n-1} e_i$.

    (b) Show that X has a spreading model isomorphic to c_0 if and only if it has a spreading model whose fundamental sequence is equivalent to the unit vector basis of c_0.

(c) Show that c_0 does not have a spreading model isomorphic to ℓ_1.

(d) Show that every spreading model over c_0 is isomorphic to c_0.

5. Prove that every spreading model on $\ell_p \oplus \ell_q$, $1 \le p \ne q < +\infty$, is isomorphic to ℓ_p or ℓ_q.

6. Show that every spreading model on the Tsirelson space T (see Chapter I, Section 7) is isomorphic to ℓ_1.

7. (a) If J is the James space defined in Exercise 5 of Chapter I, show that every sequence, $\sigma(J,J^*)$-convergent to 0 in J, has a subsequence that is equivalent to the unit vector basis of ℓ_2.

(b) Let $(e_n)_{n \in \mathbb{N}}$ be the unit vector basis of J and $s_n = \sum_{i=0}^{n-1} e_i$. Show that $(s_n)_{n \in \mathbb{N}}$ is a normalized and I.S. basic sequence of J. Prove that every bounded sequence in J with no $\sigma(J,J^*)$-convergent subsequence has a subsequence equivalent to $(s_n)_{n \in \mathbb{N}}$.

(c) What are the spreading models of J?

8. Adapt James's argument in Theorem I.4.7 to prove that X has a spreading model isomorphic to ℓ_1 (resp. c_0) if and only if, for all $\epsilon > 0$, X has a spreading model that is $(1 + \epsilon)$-isomorphic to ℓ_1 (resp. c_0).

9. Find a Banach space X and a good sequence $(x_n)_{n \in \mathbb{N}}$ in X whose spreading model is isomorphic to ℓ_1 in the following situations:

(i) $\sigma(X,X^*) - \lim_n x_n = 0$.

(ii) $\sigma(X,X^*) - \lim_n x_n = x \ne 0$.

(iii) $(x_n)_{n \in \mathbb{N}}$ is $\sigma(X,X^*)$-Cauchy, non-$\sigma(X,X^*)$-convergent.

(iv) $(x_n)_{n \in \mathbb{N}}$ is equivalent to the unit vector basis of ℓ_1.

10. (a) Let $(e_n)_{n \in \mathbb{N}}$ be an I.S. sequence such that

$$\exists \delta_1 > 0 \text{ such that } \forall k \in \mathbb{N}, \forall \epsilon_0 \cdots \epsilon_{k-1} = \pm 1,$$

$$\left\| \frac{1}{k} \sum_{i=0}^{k-1} \epsilon_i e_i \right\| \ge \delta_1$$

Prove that

$$\exists \delta_2 > 0, \text{ such that } \forall k \in \mathbb{N}, \forall a_i, \ldots, a_k \in \mathbb{R}^k,$$

$$\left\| \sum_{i=0}^{k-1} a_i e_i \right\| \ge \delta_2 \sum_{i=0}^{k-1} |a_i|$$

(b) Show that the following conditions are equivalent for an I.S.
sequence $(e_n)_{n\in\mathbb{N}}$ in a Banach space X:

    (i) $0 \in \overline{\text{conv}} \, [e_n, \, n \in \mathbb{N}]$

    (ii) $s_k = \dfrac{1}{k} \displaystyle\sum_{i=0}^{k-1} e_i \xrightarrow[k \to +\infty]{} 0$ (in norm)

    (iii) $s_k \xrightarrow[k \to +\infty]{} 0$ for $\sigma(X,X^*)$

    (iv) $\exists (s_k')_{k\in\mathbb{N}}$ subsequence of $(s_k)_{k\in\mathbb{N}}$ such that
        $s_k' \xrightarrow[k \to +\infty]{} 0$ (in norm)

    (v) The same as (iv) with $\sigma(X,X^*)$-convergence

    (vi) $e_n \xrightarrow[k \to +\infty]{} 0$ for $\sigma(X,X^*)$

    (vii) $(e_n)_{n\in\mathbb{N}}$ is basic unconditional and is not equivalent to the
        unit vector basis of ℓ_1.

11. Let $(x_n)_{n\in\mathbb{N}}$ be a good sequence in X and $(e_n)_{n\in\mathbb{N}}$ the fundamental
sequence of the spreading model defined by $(x_n)_{n\in\mathbb{N}}$.

    (a) Show that $\forall \epsilon > 0$, $\forall N \in \mathbb{N}$, $\exists v_N \in \mathbb{N}$ such that

$$\forall v_N < n_N < \cdots < n_0, \qquad \overline{\text{Span}} \, [e_i, \, i = 0, \ldots, N]$$
$$\overset{1+\epsilon}{\sim} \overline{\text{Span}} \, [x_{n_i}, \, i = 0, \ldots N]$$

    (b) Show that $(e_n)_{n\in\mathbb{N}}$ is equivalent to the unit vector basis of ℓ_1
    if and only if one of the two following conditions is satisfied:

      (i) $\exists (x_n')_{n\in\mathbb{N}} \subset (x_n)_{n\in\mathbb{N}}$, $\exists \delta > 0$, such that $\forall k \in \mathbb{N}$, $\forall \epsilon_1 \cdots \epsilon_k$
        $= \pm 1$, $\forall n_{k-1} < \cdots < n_0$,

$$\left\| \frac{1}{k} \sum_{i=0}^{k-1} \epsilon_1 x_{n_i}' \right\| \geq \delta$$

      (ii) $\exists (x_n')_{n\in\mathbb{N}} \subset (x_n)_{n\in\mathbb{N}}$, $\exists \delta > 0$, such that $\forall k \in \mathbb{N}$, $\forall k <$
        $n_{2^k-1} < \cdots < n_0$, $\forall a_0, \ldots, a_{2^k-1} \in \mathbb{R}^{2^k}$,

$$\left\| \sum_{i=0}^{2^k-1} a_i x_{n_i}' \right\| \geq \sigma \sum_{i=0}^{2^k-1} |a_i|$$

12. Let $(x_n)_{n\in\mathbb{N}}$ be a good sequence in X and $(e_n)_{n\in\mathbb{N}}$ the fundamental
sequence of the spreading model defined by $(x_n)_{n\in\mathbb{N}}$. Show that
$(e_n)_{n\in\mathbb{N}}$ is equivalent to the unit vector basis of c_0 if and only if

one of the two following conditions are satisfied:

(i) $\exists (x'_n)_{n\in\mathbb{N}} \subset (x_n)_{n\in\mathbb{N}}$, $\exists m, M > 0$ such that $\forall k \in \mathbb{N}$, $\forall k < n_{2^k-1} < \cdots < n_0$, $\forall (a_0, \ldots, a_{2^k-1}) \in \mathbb{R}^{2^k}$

$$m \sup_{0\le i\le 2^k-1} |a_i| \le \left\| \sum_{i=0}^{2^k-1} a_i x'_{n_i} \right\| \le M \sup_{0\le i\le 2^k-1} |a_i|$$

(ii) $\exists (x'_n)_{n\in\mathbb{N}} \subset (x_n)_{n\in\mathbb{N}}$, $\exists M > 0$ such that $\forall k \in \mathbb{N}$, $\forall k < n_{2^k-1} < \cdots < n_0$,

$$\left\| \sum_{i=0}^{2^k-1} x'_i \right\| \le M$$

III

Stable Banach Spaces

Stables spaces were introduced by J. L. Krivine and B. Maurey [KM] to extend a result by D. Aldous on subspaces of L_1 [A2]: Any subspace of L_1 contains some ℓ_p, almost isometrically. They proved the same statement for any stable space (Theorem III.4.1). They also proved that ℓ_p and L_p spaces are stable as well as their subspaces. So, in the class of stable Banach spaces, the problems in which we are interested, namely finding ℓ_p or c_0 subspaces and the distortion problem, are completely solved, with positive answer.

All results of this chapter were proved by J. L. Krivine and B. Maurey except the results of Section 5, which are due to J. T. Lapresté and the author [GL2].

In Section 1, spaces will always be separable.

1. DEFINITIONS, EXAMPLES, AND MAIN TOOLS

Definition III.1.1. A separable Banach space X is *stable* if for all bounded sequences $(x_n)_{n\in\mathbb{N}}$ and $(y_m)_{m\in\mathbb{N}}$ in X and all nontrivial ultrafilters $\mathcal{U}$ and $\mathcal{V}$ on $\mathbb{N}$, the following equality holds:

$$\lim_{n,\mathcal{U}} \lim_{m,\mathcal{V}} \| x_n + y_m \| = \lim_{m,\mathcal{V}} \lim_{n,\mathcal{U}} \| x_n + y_m \|$$

Observe that every subspace of a stable space is stable.

Examples

1. *Every finite-dimensional Banach space is stable*; indeed, if $(x_n)_{n \in \mathbb{N}}$ and $(y_m)_{m \in \mathbb{N}}$ are bounded in X, they converge to some x and y respectively along $\mathcal{U}$ and $\mathcal{V}$. Thus:

$$\| x + y \| = \lim_{n,\mathcal{U}} \lim_{m,\mathcal{V}} \| x_n + y_m \| = \lim_{m,\mathcal{V}} \lim_{n,\mathcal{U}} \| x_n + y_m \|$$

2. *Hilbert spaces are stable*; indeed, if $(x_n)_{n \in \mathbb{N}}$ and $(y_m)_{m \in \mathbb{N}}$ are bounded in a Hilbert space X, they converge in the $\sigma(X,X^*)$-topology to x and y, respectively, along $\mathcal{U}$ and $\mathcal{V}$. Thus, we can write

$$\lim_{n,\mathcal{U}} \lim_{m,\mathcal{V}} \| x_n + y_m \|^2 = \lim_{n,\mathcal{U}} \| x_n \|^2 + \lim_{m,\mathcal{V}} \| y_m \|^2 + 2 \lim_{n,\mathcal{U}} \lim_{m,\mathcal{V}} \langle x_n y_m \rangle$$

$$= \lim_{n,\mathcal{U}} \| x_n \|^2 + \lim_{m,\mathcal{V}} \| y_m \|^2 + 2 \langle x,y \rangle$$

and the two limits commute.

3. c_0 *is not stable*. Let $(e_n)_{n \in \mathbb{N}}$ be the unit vector basis of c_0 and set $s_m = \sum_{i=0}^{m-1} e_i$. Then

$$\begin{cases} m > n \Rightarrow \| e_n + s_m \| = 2 \\ m < n \Rightarrow \| e_n + s_m \| = 1 \end{cases}$$

and this proves our claim.

4. If X is a Banach space such that for all $\epsilon > 0$, there exists a stable Banach space X_ϵ such that $d(X,X_\epsilon) \leq 1 + \epsilon$, then X is stable.

Let us introduce the main tools of this section:

Definition III.1.2.

1. σ is a *type* on a Banach space X if there exist a bounded sequence $(x_n)_{n \in \mathbb{N}}$ in X and a nontrivial ultrafilter $\mathcal{U}$ on $\mathbb{N}$ such that

$$\forall x \in X, \qquad \sigma(x) = \lim_{n,\mathcal{U}} \| x + x_n \|$$

We will write $\sigma = ((x_n)_{n \in \mathbb{N}}, \mathcal{U})$.

σ is a *realized type* on X if there exists $x_0 \in X$ such that

$$\forall x \in X, \qquad \sigma(x) = \| x + x_0 \|$$

We will denote this type χ_{x_0}.

2. $\mathcal{T}(X)$ will denote the set of types on X. It is a subset of 1-Lipschitz functions from X to $\mathbb{R}$. We will consider $\mathcal{T}(X)$ with the topology

of *pointwise convergence*. $\hat{X}$ will denote the subset of $\mathcal{T}(X)$, composed of realized types on X. In the sequel, we will identify X and $\hat{X}$ as subsets of $\mathcal{T}(X)$. Note that, by definition, $\hat{X}$ is pointwise dense in $\mathcal{T}(X)$.

3. If σ is a type on X, namely $\sigma = ((x_n)_{n\in\mathbb{N}}, \mathcal{U}))$, we will denote

$$\| \sigma \| = \sigma(0) = \lim_{n,\mathcal{U}} \| x_n \| \quad \text{and} \quad \mathcal{T}_r(X) = \{\sigma \in \mathcal{T}(X)/\| \sigma \| \le r\}$$

for $r > 0$

Let us first give topological properties of the set of types:

Proposition III.1.3.   Let X be a separable Banach space.

(i) $\mathcal{T}(X)$ is metrizable and separable.
(ii) $\mathcal{T}_r(X)$ is compact.
(iii) $\mathcal{T}(X)$ is locally compact.

Proof

(i) If $(d_n)_{n\in\mathbb{N}}$ is a countable dense set in X, we can define a distance on $\mathcal{T}(X)$ by

$$d(\sigma,\tau) = \sum_{n=0}^{\infty} \frac{1}{2^n} \| \sigma(d_n) - \tau(d_n)\|$$

d is well defined because for all $x \in X$,

$$| \sigma(x) - \tau(x)| \le \| \sigma \| + \| \tau \|$$

It is clear that it is a distance on $\mathcal{T}(X)$ which defines the pointwise topology and that the countable set $(\chi_{d_n})_{n\in\mathbb{N}}$ is dense in $\mathcal{T}(X)$ for d.

(ii) For $\sigma \in \mathcal{T}(X)$, we can write, for all x in X,

$$\sigma(x) \le \| \sigma \| + \| x \|$$

Thus $\mathcal{T}_r(X)$ is a closed subspace of $[0, r + \| x \|]^X$; therefore $\mathcal{T}_r(X)$ is compact for the pointwise convergence.

(iii) If $\sigma_0 \in \mathcal{T}(X)$, the set $\{\sigma \in \mathcal{T}(X)/| \| \sigma \| - \| \sigma_0 \| | < \| \sigma_0 \|\}$ is an open neighborhood of σ_0 that is contained in $\mathcal{T}_{2\|\sigma_0\|}(X)$.

On $\mathcal{T}(X)$, we are going to define two operations that extend to $\mathcal{T}(X)$ the multiplication by scalars and the sum in X:

Definition III.1.4. Let X be a separable stable Banach space. Let $\sigma = ((x_n)_{n\in\mathbb{N}}, \mathcal{U})$ and $\tau = ((y_m)_{m\in\mathbb{N}}, \mathcal{V})$ belong to $\mathcal{T}(X)$.

The type $\lambda\sigma$ for $\lambda \in \mathbb{R}$ is defined by

$$\forall x \in X, \qquad \lambda\sigma(x) = \lim_{n,\mathcal{U}} \| x + \lambda x_n \|$$

The convolution product $\sigma * \tau$ is defined by

$$\forall x \in X, \qquad \sigma * \tau(x) = \lim_{n,\mathcal{U}} \lim_{m,\mathcal{V}} \| x + x_n + y_m \|$$

Remark. This last definition makes sense only if we verify that $\sigma * \tau$ does not depend on sequences that define σ and τ. Suppose that

$$\sigma = ((x_n)_{n\in\mathbb{N}}, \mathcal{U}) = ((x'_n)_{n\in\mathbb{N}}, \mathcal{U}')$$

$$\tau = ((y_m)_{m\in\mathbb{N}}, \mathcal{V}) = ((y'_m)_{m\in\mathbb{N}}, \mathcal{V}')$$

Then, for all $x \in X$, we can write

$$\lim_{n,\mathcal{U}'} \lim_{m,\mathcal{V}'} \| x + x'_n + y'_m \| = \lim_{n,\mathcal{U}'} \tau(x + x'_n)$$

$$= \lim_{n,\mathcal{U}'} \lim_{m,\mathcal{V}} \| x + x'_n + y_m \|$$

$$= \lim_{m,\mathcal{V}} \lim_{n,\mathcal{U}'} \| x + x'_n + y_m \|$$

$$= \lim_{m,\mathcal{V}} \sigma(x + y_m)$$

$$= \lim_{m,\mathcal{V}} \lim_{n,\mathcal{U}} \| x + x_n + y_m \| = \sigma * \tau(x)$$

So Definition III.1.4 makes sense and it is clear that $\sigma * \tau$ belongs to $\mathcal{T}(X)$.

The next result shows some very important properties of the convolution product of types.

Proposition III.1.5. Let X be a stable Banach space and $\sigma, \tau \in \mathcal{T}(X)$. Then

(i) $\sigma * \tau = \tau * \sigma$

(ii) The map

$$\phi : \mathcal{T}(X) \times \mathcal{T}(X) \to \mathbb{R}$$

$$(\sigma, \tau) \mapsto \sigma * \tau(0)$$

is a separately continuous function.

Proof

    (i) is obvious by definition

    (ii) Fix $\tau \in \mathcal{T}(X)$. We want to show that the map:

$$\psi : \mathcal{T}(X) \to \mathbb{R} \qquad \text{is continuous}$$

$$\sigma \mapsto \sigma * \tau(0)$$

If $(\chi_{x_n})_{n \in \mathbb{N}}$ is a sequence of realized types, converging pointwise to σ, we can write

$$\forall x \in X, \qquad \sigma(x) = \lim_n \| x + x_n \| = \lim_n \chi_{x_n}(x)$$

Thus, if $\tau = ((y_m)_{m \in \mathbb{N}}, \mathcal{V})$, we get

$$\psi(\sigma) = \sigma * \tau(0) = \lim_n \lim_{m, \mathcal{V}} \| x_n + y_m \|$$

$$= \lim_n \tau(x_n) = \lim_n \tau * \chi_{x_n}(0)$$

$$= \lim_n \psi(\chi_{x_n})$$

Therefore, $(\psi(\chi_{x_n}))_{n \in \mathbb{N}}$ converges to $\psi(\sigma)$. Since $\hat{X}$ is dense in $\mathcal{T}(X)$, an easy diagonal process shows that if $(\sigma_n)_{n \in \mathbb{N}}$ is a sequence of non-necessarily realized types converging to σ, then $(\psi(\sigma_n))_{n \in \mathbb{N}}$ converges to $\psi(\sigma)$. Since $\mathcal{T}(X)$ is metrizable, this proves our claim.

Remark III.1.6.   It is easy to deduce from Proposition III.1.5 that, for all $x \in X$, $(\sigma, \tau) \to \sigma * \tau(x)$ is also separately continuous. Thus if $(\sigma_n)_{n \in \mathbb{N}}$ converges to σ pointwise, then for all $\tau \in \mathcal{T}(X)$, $(\sigma_n * \tau)_{n \in \mathbb{N}}$ converges to $\sigma * \tau$ pointwise.

2. CHARACTERIZATIONS OF STABLE BANACH SPACES

In this section we are going to give two representations of the function $\| x + y \|$ in a stable space; these representations will allow us to prove that ℓ_p and L_p spaces are stable (Corollary III.2.7).

Theorem III.2.1.   Fix $p \in [1, +\infty[$. A Banach space X is stable if and only if there exist a reflexive space $\mathcal{E}$, a dense set B in B_X, the unit ball of X, and two bounded maps

$$U : B \to \mathscr{E} \quad \text{and} \quad V : B \to \mathscr{E}*$$

such that, for all x and y in B_X,

$$\| x + y \|^p = \langle Ux, Vy \rangle$$

Remark. In this setting, maps U and V do not need to be linear.

To prove this theorem, we need to study separately continuous functions of two variables to be able to use Proposition III.1.5.

Proposition III.2.2. Let $\phi : K \times K' \to \mathbb{R}$ be a bounded and separately continuous function on two metrizable compact sets K and K'. There exist a reflexive Banach space $\mathscr{E}$ and two bounded maps $U : K \to \mathscr{E}$ and $V : K' \to \mathscr{E}*$ such that, for all x in K and y in K', $\phi(x,y) = \langle Ux, Vy \rangle$.

Before starting the proof of this proposition, we need a technical lemma:

Lemma III.2.3. Under the hypothesis of Proposition III.2.2, ϕ is of the first Baire class.

Proof of Lemma III.2.3. Recall that ϕ is of the *first Baire class* if it is a pointwise limit of a sequence of continuous functions. Let d be a distance on K. For $a \in K$ and $n \in \mathbb{N}$, we define a function ξ_a^n on K by

$$\forall x \in K, \qquad \xi_a^n(x) = \mathrm{Sup} \left\{ \frac{1}{n} - d(a,x), 0 \right\}$$

ξ_a^n is continuous on K, strictly positive if $d(a,x) \le 1/2n$, and null if $d(a,x) \ge 1/n$.

Let $(a_i^n)_{i \in I_n}$ be a $1/2n$-net in K (see Definition I.1.8) and define a function f_i^n on K and a function ϕ_n on $K \times K'$ by

$$\begin{cases} \forall x \in K, & f_i^n(x) = \dfrac{\xi_{a_i^n}^n(x)}{\displaystyle\sum_{j \in I_n} \xi_{a_j^n}^n(x)} \\[2ex] \forall x,y \in K \times K', & \phi_n(x,y) = \displaystyle\sum_{i \in I_n} \phi(a_i^n,y) f_i^n(x) \end{cases}$$

Note that $\sum_{i \in I_n} f_i^n$ is function 1 on K and that ϕ_n is continuous on $K \times K'$. Let us show that $(\phi_n)_{n \in \mathbb{N}}$ converges pointwise to ϕ:

Fix $(x,y) \in K \times K'$ and $\epsilon > 0$. Then for all $n \in \mathbb{N}$, there exists i_n such that $d(x,a_{i_n}^n) \leq 1/2n$ and there exists $N \in \mathbb{N}$ such that

$$n \geq N \Rightarrow |\phi(x,y) - \phi(a_{i_n}^n,y)| \leq \epsilon$$

Thus we can write, for $n \geq N$,

$$|\phi(x,y) - \phi_n(x,y) = |\sum_{i \in I_n} [\phi(x,y) - \phi(a_i^n,y)] f_i^n(x)|$$

$$\leq \sum_{i \in I_n} |\phi(x,y) - \phi(a_i^n - y)| f_i^n(x)$$

$$\leq |\phi(x,y) - \phi(a_{i_n}^n,y)| f_{i_n}^n(x) \leq \epsilon f_{i_n}^n(x) \leq \epsilon$$

This proves that $(\phi_n)_{n \in \mathbb{N}}$ converges pointwise to ϕ and thus that ϕ is of the first Baire class.

Proof of Proposition III.2.2. Define an operator $T : [C(K)]^* \to C(K')$ by

$$\forall \mu \in [C(K)]^*, \forall y \in K', \qquad (T\mu)(y) = \int_K \phi(x,y)\mu(dx)$$

Lebesgue's theorem implies that $T\mu$ belongs to $C(K')$. Since ϕ is of the first Baire class, it is borelian and by Fubini's theorem, we get

$$\forall v \in [C(K')]^*, \qquad \langle T\mu,v \rangle = \int_{K'} \int_K \phi(x,y)\,d\mu(x)\,dv(y) = \langle \mu,T^*v \rangle$$

with $T^*v(x) = \int_{K'} \phi(x,y)v(dy)$. Thus T^* maps $[C(K')]^*$ into $C(K)$. Let us show that this implies that T is weakly compact.

Recall that an operator from X to Y is *weakly compact* (see [DS]) if the image of the unit ball of X is relatively $\sigma(Y,Y^*)$-compact in Y. Indeed, by Eberlein-Smulian's theorem, to prove that the image of the unit ball of $[C(K)]^*$ by T is relatively $\sigma(C(K'),C(K')^*)$-compact in $C(K')$, it is sufficient to show that every sequence $(g_n)_{n \in \mathbb{N}}$ in $T(B_{C(K)^*})$ has a $\sigma(C(K'),C(K')^*)$-convergent subsequence.

By definition, for all $n \in \mathbb{N}$, there exists μ_n in $B_{C(K)^*}$ such that

$$g_n = T(\mu_n)$$

Let v belong to $C(K')^*$. Then by definition of T^* we have

$$(v,g_n) = (v,T\mu_n) = (T^*v,\mu_n)$$

We know that T^*v belongs to $C(K)$; thus, since $B_{C(K)^*}$ is relatively $\sigma(C(K)^*,C(K))$-compact, there exists a subsequence $(\mu_{n_k})_{k\in\mathbb{N}}$ of $(\mu_n)_{n\in\mathbb{N}}$ which is $\sigma(C(K)^*,C(K))$-convergent. And this proves that $(g_{n_k})_{k\in\mathbb{N}}$ is $\sigma(C(K'),C(K')^*)$-convergent.

We can now apply a result of W. J. Davis, T. Figiel, W. B. Johnson, and A. Pelczynski [DFJP]:

Every weakly compact operator factors through a reflexive space.

Thus there exist a reflexive Banach space $\mathscr{E}$ and continuous operators $j : [C(K)]^* \to \mathscr{E}$ and $h : \mathscr{E} \to C(K')$ such that $T = hj$. For $x \in K$, $y \in K'$, let δ_x and δ_y be the corresponding Dirac measures on K and K', respectively. We get

$$\phi(x,y) = \langle T\delta_x, \delta_y \rangle = \langle hj\,\delta_x, \delta_y \rangle = \langle j\delta_x, h^*\delta_y \rangle$$

Set

$$\begin{cases} \forall x \in K, & Ux = j\delta_x \\ \forall y \in K', & Vy = h^*\delta_y \end{cases}$$

Then U and V verify the conclusion of Proposition III.2.2.

We can now prove the main result:

Proof of Theorem III.2.1. If X is a stable Banach space, to prove the existence of $\mathscr{E}$, U, and V, we apply Proposition III.2.2 with

$$K = K' = \mathscr{T}_1(X) \quad \text{and} \quad \phi(\sigma,\tau) = \|\sigma * \tau\|^p$$

We get a reflexive space $\mathscr{E}$ and two maps U and V from $\mathscr{T}_1(X)$ to $\mathscr{E}$ and $\mathscr{E}'$ respectively such that if $(x,y) \in B_X^2$, by taking $\sigma = \chi_x$ and $\tau = \chi_y$,

$$\|x + y\|^p = \langle Ux, Vy \rangle$$

On the other hand, suppose that $\mathscr{E}$, B, U, and V are given. If $(x_n)_{n\in\mathbb{N}}$ and $(y_m)_{m\in\mathbb{N}}$ are bounded sequences in X, and $\mathscr{U}$ and $\mathscr{V}$ are two ultrafilters on $\mathbb{N}$, define

$$M = \sup_{n\in\mathbb{N}} \|x_n\|, \qquad N = \sup_{n\in\mathbb{N}} \|y_m\|, \qquad P = \sup\{M,N\}$$

$$x'_n = \frac{x_n}{P}, \qquad y'_m = \frac{y_m}{P} \quad \text{for all } n,m \in \mathbb{N}^2$$

Then $(x'_n)_{n\in\mathbb{N}}$ and $y'_m)_{m\in\mathbb{N}}$ are in B_X. Since B is dense in B_X, for all n and m in $\mathbb{N}$, there exist $x''_n, y''_n \in B$ such that

$$\| x'_n - x''_n \| \leq \frac{1}{n} \quad \text{and} \quad \| y'_m - y''_m \| \leq \frac{1}{m}$$

Since $\mathscr{E}$ is reflexive and U and V are bounded, $(Ux''_n)_{n\in\mathbb{N}}$ has a $\sigma(\mathscr{E}, \mathscr{E}^*)$-limit α along $\mathscr{U}$ and $(Vy''_m)_{m\in\mathbb{N}}$ has a $\sigma(\mathscr{E}^*, \mathscr{E})$-limit β along $\mathscr{V}$. So we get

$$\lim_{n,\mathscr{U}} \lim_{m,\mathscr{V}} \| x_n + y_m \|^p = P \lim_{n,\mathscr{U}} \lim_{m,\mathscr{V}} \| x'_n + y'_m \|^p$$

$$= P \lim_{n,\mathscr{U}} \lim_{m,\mathscr{V}} \| x''_n + y''_m \|^p$$

$$= P \lim_{n,\mathscr{U}} \lim_{m,\mathscr{V}} \langle Ux''_n, Vy''_m \rangle = P \langle \alpha, \beta \rangle$$

If we change the order of the limits, this computation would give the same result and thus X is stable.

This representation of the norm of stable spaces is not sufficient; indeed, the function $\| x + y \|^p$ is represented by a reflexive space only if x and y are in a dense set of the unit ball. The next result gives a more complicated representation of this function but on the whole space:

Theorem III.2.4. Fix $1 \leq p < +\infty$. The Banach space X is stable if and only if there exist $k \in \mathbb{N}$, $c > 0$ and for all $i \in \{0, \ldots, k\}$, $0 < \alpha_i < p$, $\mathscr{E}_i$ reflexive spaces, $\mathbb{U}_i : X \to \mathscr{E}_i$, and $\mathbb{V}_i : X \to \mathscr{E}_i^*$ bounded maps such that

$$\forall(x,y) \in X^2, \quad \| x + y \|^p - \| x \|^p - \| y \|^p = \sum_{i=0}^{k} \langle \mathbb{U}_i x, \mathbb{V}_i y \rangle$$

$$\| \mathbb{U}_i x \|_{\mathscr{E}_i} \leq c \| x \|^{\alpha_i} \quad \text{for } i = 0, \ldots, k$$

$$\| \mathbb{V}_i y \|_{\mathscr{E}_i^*} \leq c \| y \|^{p - \alpha_i} \quad \text{for } i = 0, \ldots, k$$

where $\langle \, , \, \rangle$ denotes the duality product between $\mathscr{E}_i$ and $\mathscr{E}_i^*$.

We need two lemmas to prove this theorem. Recall first that the *Alexandroff compactification* of a locally compact set K is a compact set $\overline{K}$ that is equal to the union of K and a point $\{\infty\}$.

The existence and properties of $\overline{K}$ were proved by Alexandroff; see [Bo1].

Lemma III.2.5. Fix $1 \leq p < 2$. Let X be a stable space and denote by $\overline{\mathcal{T}(X)}$ the Alexandroff compactification of $\mathcal{T}(X)$. Then there exists $\varphi : X \times X \to \mathbb{R}$ with a bounded and separately continuous extension to $\overline{\mathcal{T}(X)} \times \overline{\mathcal{T}(X)}$ such that

$$\forall x,y \in X^2, \qquad \| x + y \|^p - \| x \|^p - \| y \|^p = (\| x \|^{p/2} \| y \|^{p/2}) \, \varphi(x,y)$$

Proof of Lemma III.2.5. Since $1 \leq p < 2$, it is easy to show that there exists $c > 0$ such that for all $(s,t) \in \mathbb{R}^2$:

$$\| t + s \|^p - | t |^p - | s |^p \| \leq c(| s | | t |^{p-1} \wedge | t | | s |^{p-1})$$

Thus, if $\| x + y \|^p - \| x \|^p - \| y \|^p \geq 0$, we can apply this inequality with $t = \| x \|$ and $s = \| y \|$ and we get

$$0 \leq \| x + y \|^p - \| x \|^p - \| y \|^p \leq (\| x \| + \| y \|)^p - \| x \|^p - \| y \|^p$$

$$\leq c(\| y \| \| x \|^{p-1} \wedge \| x \| \| y \|^{p-1})$$

If $\| x + y \|^p - \| x \|^p - \| y \|^p \leq 0$, we take $t = \| x \|$ and $s = - \| y \|$ and get

$$0 \leq \| x \|^p + \| y \|^p - \| x + y \|^p \leq \| x \|^p + \| y \|^p - | \| x \| - \| y \| |^p$$

$$\leq c(\| x \| \| y \|^{p-1} \wedge \| y \| \| x \|^{p-1})$$

So, if we define $\varphi(x,y)$, for x and $y \neq 0$ by,

$$\| x + y \|^p - \| x \|^p - \| y \|^p - \| x \|^{p/2} \| y \|^{p/2} \, \varphi(x,y)$$

we have

$$| \varphi(x,y) | \leq c \left[\left(\frac{\| x \|}{\| y \|} \right)^{1-p/2} \wedge \left(\frac{\| y \|}{\| x \|} \right)^{1-p/2} \right] \leq c$$

This function φ can be extended to a bounded function on $\mathcal{T}(X) - \{0\}$ by setting, for σ and τ in $\mathcal{T}(X) - \{0\}$

$$\| \sigma * \tau \|^p - \| \sigma \|^p - \| \tau \|^p = \| \sigma \|^{p/2} \| \tau \|^{p/2} \, \varphi(\sigma,\tau)$$

If σ or τ is 0 or $+\infty$, we define $\varphi(\sigma,\tau) = 0$.
Thus, φ is a bounded function on $\overline{\mathcal{T}(X)}$ which is obviously separately continuous on $\mathcal{T}(x) - \{0\}$.

The inequality

$$| \varphi(x,y) | \leq c \left[\left(\frac{\| x \|}{\| y \|} \right)^{1-p/2} \wedge \left(\frac{\| y \|}{\| x \|} \right)^{1-p/2} \right]$$

can be extended to types by the same formula, namely

$$| \varphi(\sigma,\tau) | \leq c \left[\left(\frac{\| \sigma \|}{\| \tau \|} \right)^{1-p/2} \wedge \left(\frac{\| \tau \|}{\| \sigma \|} \right)^{1-p/2} \right]$$

This proves that φ is also separately continuous at 0 and $+\infty$.

Lemma III.2.6. Fix $1 \leq p < +\infty$. If X is a stable space, there exist $k \in \mathbb{N}$, and for all $i \in \{0, \ldots, k\}$, real numbers α_i with $0 < \alpha_i < p$ and maps φ_i from $X \times X$ into $\mathbb{R}$ with bounded and separately continuous extensions to $\overline{\mathcal{T}(X)} \times \overline{\mathcal{T}(X)}$ such that, for all x and y in X,

$$\| x + y \|^p - \| x \|^p - \| y \|^p + \sum_{i=0}^{k} \| x \|^{\alpha_i} \| y \|^{p-\alpha_i} \varphi_i(x,y)$$

Proof of Lemma III.2.6. If $1 \leq p < 2$, Lemma III.2.5 gives the result with $k = 0$, $\alpha_0 = p/2$, $\varphi_0 = \varphi$.

To prove this result, it suffices to show that if it is true for p, then it is also true for $2p$. Let us suppose that for x, y in X^2, we have

$$\| x + y \|^p = \| x \|^p + \| y \|^p + \sum_{i=0}^{k} \| x \|^{\alpha_i} \| y \|^{p-\alpha_i} \varphi_i(x,y)$$

Taking the square of both sides, we get

$$\| x + y \|^{2P} = \| x \|^{2p} + \| y \|^{2p} + S(x,y)$$

where S is a sum of terms of the desired form.

Proof of Theorem III.2.4. The "if" part is obvious.

Suppose that X is stable and apply Lemma III.2.6. We get the existence of maps φ_i and scalars α_i such that

$$\| x + y \|^p - \| x \|^p - \| y \|^p = \sum_{i=0}^{k} \| x \|^{\alpha_i} \| y \|^{p-\alpha_i} \varphi_i(x,y)$$

Then, for all $i = 0, \ldots, k$, we can apply Proposition III.2.2 to each φ_i on $\overline{\mathcal{T}(x)} \times \overline{\mathcal{T}(X)}$: we get the existence of bounded maps:

$$U_i : \overline{\mathcal{T}(X)} \to \mathscr{E} \quad \text{and} \quad V_i : \overline{\mathcal{T}(X)} \to \mathscr{E}^*$$

such that for all x and y in X,

$$\varphi(x,y) = \langle U_i x, V_i y \rangle$$

Define U_i and V_i by

$$\begin{cases} U_i x = \| x \|^{\alpha_i} U_i x \\ V_i y = \| y \|^{p - \alpha_i} V_i y \end{cases}$$

Then the conclusion of the theorem holds with

$$c = \operatorname{Sup} \{ \| U_i \sigma \|, \| V_i \tau \| / i = \{0, \ldots, k\}, (\sigma,\tau) \in \overline{\mathcal{T}(X)^2} \}$$

Remark. This representaton of the function $\| x + y \|^p - \| x \|^p - \| y \|^p$ on $X \times X$ is also valid for the function $\| \sigma * \tau \|^p - \| \sigma \|^p - \| \tau \|^p$ on $\overline{\mathcal{T}(X)} \times \overline{\mathcal{T}(X)}$.

As a consequence of this property of the norm of stable spaces, we have the following announced result:

Corollary III.2.7. Let (Ω,Σ,μ) be a measured space and $L_p = L_p(\Omega,\Sigma,\mu)$ for $1 \le p < +\infty$. If X is a (separable) stable space, then $L_p(X) = L_p(\Omega,\Sigma,\mu;X)$ is stable. In particular, L_p and ℓ_p are stable.

Proof. Let $(f_n)_{n \in \mathbb{N}}$ and $(g_m)_{m \in \mathbb{N}}$ be two bounded sequences in $L_p(X)$ and let $\mathcal{U}$ and $\mathcal{V}$ be two ultrafilters on $\mathbb{N}$. Since X is stable, with the notation of Theorem III.2.4, we can write, for almost all $\omega \in \Omega$,

$$\begin{cases} \| f_n(\omega) + g_m(\omega) \|_X^p = \| f_n(\omega) \|_X^p + \| g_m(\omega) \|_X^p + \displaystyle\sum_{i=0}^{k} \langle U_i(\omega), V_i(\omega) \rangle \\ \| U_i f_n(\omega) \|_{\mathcal{E}_i} \le c \| f_n(\omega) \|_X^{\alpha_i} & \text{for } i = 0, \ldots, k \\ \| V_i g_m(\omega) \|_{\mathcal{E}_i^*} \le c \| g_m(\omega) \|_X^{p - \alpha_i} & \text{for } i = 0, \ldots, k \end{cases}$$

Then we get

$$\begin{cases} \displaystyle\int \| U_i f_n(\omega) \|_{\mathcal{E}_i}^{p/\alpha_i} d\mu(\omega) \le C^{p/\alpha_i} \int \| f_n(\omega) \|_X^p \, d\mu(\omega) \\ \qquad\qquad = C^{p/\alpha_i} \| f_n \|_{L_p(x)}^p \\ \displaystyle\int \| V_i g_m(\omega) \|_{\mathcal{E}_i^*}^{p/(p - \alpha_i)} d\mu(\omega) \le C^{p/(p - \alpha_i)} \int \| g_m(\omega) \|_X^p \, d\mu(\omega) \\ \qquad\qquad = C^{p/(p - \alpha_i)} \| g_m \|_{L_p(X)}^p \end{cases}$$

Since $L_{p/(p-\alpha_i)}$ $(\mathscr{E}_i^*)$ is the dual of L_{p/α_i} $(\mathscr{E}_i)$ and since these spaces are reflexive (see [DU]), sequences $(\mathbb{U}_i f_n)_{n\in\mathbb{N}}$ and $(\mathbb{V}_i g_m)_{m\in\mathbb{N}}$ have $\sigma(L_{p/\alpha_i}(\mathscr{E}_i), L_{p/p-\alpha_i}(\mathscr{E}_i^*))$ and $\sigma(L_{p/p-\alpha_i}(\mathscr{E}_i^*), L_{p/\alpha_i}(\mathscr{E}_i))$ limits F_i and G_i along $\mathscr{U}$ and $\mathscr{V}$, respectively, for all $i = 0, \ldots, k$.

So, we can write

$$\lim_{n,\mathscr{U}} \lim_{m,\mathscr{V}} \| f_n + g_m \|^p_{L_p(X)} = \lim_{n,\mathscr{U}} \| f_n \|^p_{L_p(X)}$$

$$+ \lim_{m,\mathscr{V}} \| g_m \|^p_{L_p(X)} + \sum_{i=0}^{k} \langle F_i, G_i \rangle$$

where here $\langle \, , \, \rangle$ also denotes the duality product between $L_{p/\alpha_i}(\mathscr{E}_i)$ and $L_{p/p-\alpha_i}(\mathscr{E}_i^*)$, that is, $\langle F, G \rangle = \int \langle F(\omega), G(\omega) \rangle \, d\mu(\omega)$.

It is clear that this result does not depend on the order of the limits and this proves Corollary III.2.7.

If we take $X = \mathbb{R}$, then we obtain that any L_p-space is stable for all p with $1 \leq p < +\infty$. In particular ℓ_p and $L_p[0,1]$ are stable.

3. SPREADING MODELS ON STABLE SPACES

There is a natural correspondence between spreading models and types on a stable space that is given by the following computation:

Let X be a stable space and $(x_n)_{n\in\mathbb{N}}$ a bounded sequence in X, with no convergent subsequence. If σ is the type on X defined by $(x_n)_{n\in\mathbb{N}}$ and a ultrafilter $\mathscr{U}$ and if $(e_n)_{n\in\mathbb{N}}$ is the fundamental sequence of the spreading model defined by $(x_n)_{n\in\mathbb{N}}$ and $\mathscr{U}$, we get, for all $x \in X$, $k \in \mathbb{N}$, and $(a_i)_{i=0,\ldots,k} \in \mathbb{R}^{k+1}$,

$$(*) \quad a_0\sigma * \cdots * a_k\alpha(x) = \lim_{n_0,\mathscr{U}} \cdots \lim_{n_k,\mathscr{U}} \| x + a_0 x_{n_0} + \cdots + a_k x_{n_k} \|$$

$$= \lim_{n_k,\mathscr{U}} \cdots \lim_{n_0,\mathscr{U}} \| x + a_0 x_{n_0} + \cdots + a_k x_{n_k} \|$$

$$= \| x + a_0 e_0 + \cdots + a_k e_k \|$$

Remark. In general, there are several types associated with a given sequence $(x_n)_{n\in\mathbb{N}}$; it depends on the ultrafilter $\mathscr{U}$. However, as noted in Remark II.2.8 for spreading models on good sequences, if $(x_n)_{n\in\mathbb{N}}$ is a good sequence in X, there is only one type σ, defined by $(x_n)_{n\in\mathbb{N}}$;

indeed $\lim_n \| x + x_n \|$ exists for all $x \in X$ and then we do not have to specify any ultrafilter $\mathcal{U}$.

This correspondence between types and spreading models on stable spaces gives further properties of fundamental sequences of spreading models. Let us first give definitions:

Definition III.3.1

1. A sequence $(y_n)_{n \in \mathbb{N}}$ in a Banach space X is said to be *K-exchangeable* if for all permutations π on $\mathbb{N}$, $(y_{\pi(n)})_{n \in \mathbb{N}}$ is K-equivalent to $(y_n)_{n \in \mathbb{N}}$.
2. $(y_n)_{n \in \mathbb{N}}$ is called *K-symmetric* if for all permutation π and all $(\epsilon_n)_{n \in \mathbb{N}}$ with $\epsilon_n = \pm 1$ for all $n \in \mathbb{N}$, $(\epsilon_n y_{\pi(n)})_{n \in \mathbb{N}}$ is K-equivalent to $(y_n)_{n \in \mathbb{N}}$.

These two definitions have obvious analogous formulations over the space, as was done in II.2.1 or II.2.3.

Remark III.3.2. A basic sequence $(y_n)_{n \in \mathbb{N}}$ with basic constant K that is K'-exchangeable is KK'-unconditional and $2KK'$-sign-unconditional (see Definitions I.3.4 and II.5.2). An obvious analogous statement is true over the space.

Proposition III.3.3. Let $(e_n)_{n \in \mathbb{N}}$ be the fundamental sequence of a spreading model on a stable space X. Then $(e_n)_{n \in \mathbb{N}}$ is 1-exchangeable over X.

Proof. If σ is the type associated with $(e_n)_{n \in \mathbb{N}}$ by (*) we get, for all x in X, k in $\mathbb{N}$, $(a_0, \ldots, a_k)$ in $\mathbb{R}^{k+1}$ and all permutation π on $\mathbb{N}$,

$$\| x + \sum_{i=0}^{k} a_i e_{\pi(i)} \| = a_0 \sigma * \cdots * a_k \sigma(x) = \| x + \sum_{i=0}^{k} a_i e_i \|$$

Definition III.3.4

1. A type σ on X is said to be *symmetric* if $\forall x \in X$, $\sigma(x) = \sigma(-x)$.
2. The subset of the *symmetric types* of X is denoted by $\mathcal{S}(X)$ and we set $\mathcal{S}_r(X) = \{\sigma \in \mathcal{S}(X), \| \sigma \| \leq r\}$.

Remark III.3.5. If σ is a realized symmetric type, it is the null type.

If X is infinite-dimensional, there exists a nonnull symmetric type on X; indeed, if σ is a nonrealized type on X, then $\sigma^*(-\sigma)$ is symmetric.

Proposition III.3.6. Let $(e_n)_{n \in \mathbb{N}}$ be the fundamental sequence of a spreading model on a stable space X and σ the type on X, associated with $(e_n)_{n \in \mathbb{N}}$ by (*). Then σ is symmetric, if and only if $(e_n)_{n \in \mathbb{N}}$ is 1-sign-unconditional over X.

Proof. We can write, for all $x \in X$, $k \in \mathbb{N}$, $(a_i)_{i=0,\ldots,k} \in \mathbb{R}^{k+1}$, and $(\epsilon_i)_{i=0,\ldots,k}$, with $\epsilon_i = \pm 1$ for all $i = 0, \ldots, k$,

$$\| x + \sum_{i=0}^{k} \epsilon_i a_i e_i \| = a_0 \sigma * \cdots * a_k \sigma(x) = \| x + \sum_{i=0}^{k} a_i e_i \|$$

4. THE EXISTENCE OF ℓ_p SUBSPACES

This is the main property of stable spaces. It is due to J. L. Krivine and B. Maurey [KM] for stable spaces and it was proved earlier by D. Aldous in [A2] for subspaces of L_1.

Theorem III.4.1. Let X be a stable Banach space. Then there exists $p \in [1, +\infty[$ such that for all $\epsilon > 0$, there is a subspace of X that is $(1 + \epsilon)$-isomorphic to ℓ_p.

To prove this theorem, we need some definitions and lemmas:

Definition III.4.2

1. A symmetric type σ on a stable space X is a c_0-type or an ℓ_p-type ($1 \le p < +\infty$) if, for all $\alpha, \beta \ge 0$, we have

$$\alpha\sigma * \beta\sigma = \mathrm{Sup}(\alpha,\beta)\, \sigma \quad \text{or} \quad (\alpha^p + \beta^p)^{1/p}\, \sigma$$

2. C is a *conic class* of $\mathcal{S}(X)$ if

   (a) $C \ne \{0\}$
   (b) If σ belongs to C then for all $\lambda \ge 0$, $\lambda\sigma$ belongs to C
   (c) If σ and τ belong to C then $\sigma * \tau$ belongs to C

3. If C is a conic class, we define $C_1 = C \cap \mathcal{T}_1(X)$.

Remarks III.4.3

1. If σ is a c_0-type or an ℓ_p-type $(1 \leq p < +\infty)$ on X, then the spreading model associated with σ by (*) is isometric to c_0 or ℓ_p over X and by Theorem II.4.2, for all $\epsilon > 0$, X contains a subspace $(1 + \epsilon)$-isomorphic to c_0 or ℓ_p.

2. If X is stable, since c_0 is not stable, there is no c_0-type on X.

Lemma III.4.4. Every closed conic class contains a minimal closed conic class.

Proof of Lemma III.4.4. The set of all conic classes is ordered by inclusion. If $\mathscr{C}$ is a totally ordered family of closed conic classes, then $\cap_{C \in \mathscr{C}} C$ verifies (b) and (c). Moreover, every conic class C contains C_1, which is a compact set in $\mathscr{T}(X)$. Thus $\cap_{C \in \mathscr{C}} C \supset \cap_{C \in \mathscr{C}} C_1 \neq \{0\}$ and this set also satisfies (a). Then $\cap_{C \in \mathscr{C}} C$ is a conic class and the result of the lemma is a simple consequence of Zorn's axiom.

Lemma III.4.5. Let C be a closed conic class of $\mathscr{S}(X)$. Then there exists $p \in [1, +\infty]$ such that for all $\epsilon > 0$ and all $n \in \mathbb{N}$, there exists $\sigma^n \in C_1$ such that, for all $(a_i)_{i=0,\ldots,n-1} \in \mathbb{R}^n$ and $x \in X$:

$$a_0 \sigma^n * \cdots * a_{n-1} \sigma^n(x) \overset{(1 \pm \epsilon)}{\sim} \left(\sum_{i=0}^{n-1} |a_i|^p \right)^{1/p} \sigma^n(x) \quad \text{if } p < +\infty$$

$$\text{or} \qquad \text{Sup} \, |a_i| \, \sigma^n(x) \qquad \text{if } p = +\infty$$

Proof of Lemma III.4.5. Let $\sigma \in C$ and $(x_i)_{i \in \mathbb{N}}$ be a good sequence in X that defines σ. If $(e_i)_{i \in \mathbb{N}}$ is the fundamental sequence of the spreading model defined by $(x_i)_{i \in \mathbb{N}}$, $(e_i)_{i \in \mathbb{N}}$ is 1-sign-unconditional by Proposition III.3.6. Then the proof of Theorem II.5.1 shows that there exist $p \in [1, +\infty]$ and for all $n \in \mathbb{N}$ and $\epsilon > 0$, normalized blocks $(h_i)_{i=0,\ldots,n-1}$ on $(e_i)_{i \in \mathbb{N}}$ such that, for all x in X and $(a_i)_{i=0,\ldots,n-1}$ in $\mathbb{R}^n$:

$$(1 - \epsilon) \left\| x + \left(\sum_{i=0}^{n-1} |a_i|^p \right)^{1/p} h_1 \right\| \leq \left\| x + \sum_{i=0}^{n-1} a_i h_i \right\|$$

$$\leq (1 + \epsilon) \left\| x + \left(\sum_{i+0}^{n-1} |a_i|^p \right)^{1/p} h_1 \right\|$$

(If $p = +\infty$, replace

$$\left\| x + \left(\sum_{i+0}^{n-1} |a_i|^p \right)^{1/p} h_1 \right\|$$

by $\| x + \mathrm{Sup}_{i=0,\ldots,n-1} |a_i| h_1 \|$).

Moreover, Remark II.5.14 shows that these blocks $h_0, \ldots, h_{n-1}$ have the same coefficients on $(e_i)_{i\in\mathbb{N}}$, say

$$h_i = \sum_{j=0}^{N-1} \alpha_j e_{j+Ni} \quad \text{for } i = 0, \ldots, n-1$$

Thus, each h_i defines on X the same type:

$$\sigma^n = \alpha_0 \sigma * \cdots * \alpha_{N-1} \sigma$$

Obviously σ^n belongs to C_1 and has the desired property.

Definition III.4.6.   A sequence of types $(\sigma^n)_{n\in\mathbb{N}}$ is said to be *p-approximating* if

(a) $\underset{n\in\mathbb{N}}{\mathrm{Sup}} \, \| \sigma^n \| < +\infty$

(b)   There exists a sequence $(\epsilon_n)_{n\in\mathbb{N}}$ of positive numbers converging to 0 such that, for all $n \in \mathbb{N}$ and $(a_i)_{i=0,\ldots,n-1} \in \mathbb{R}^n$,

$$a_0\alpha^n * \cdots * a_{n-1}\sigma^n(n) \overset{(1+\epsilon_n)}{\underset{\sim}{}} \left(\sum_{i=0}^{n-1} |a_i|^n \right)^{1/p} \sigma^n(x) \quad \text{if } p < +\infty$$

$$\text{or} \qquad \underset{i=0,\ldots,n-1}{\mathrm{Sup}} |a_i| \, \sigma^n(x) \quad \text{if } p = +\infty$$

Lemma III.4.7.   Let C be a minimal closed conic class. Then there exists $p \in [1, +\infty]$ such that every type belonging to C is the limit of a p-approximating sequence.

Proof of Lemma III.4.7.   Let C be a closed minimal conic class and $(\epsilon_n)_{n\in\mathbb{N}}$ a sequence of positive numbers converging to 0. For all $n \in \mathbb{N}$, denote by σ^n the type obtained by Lemma III.4.5, applied to ϵ_n and n. We obtain a sequence $(\sigma^n)_{n\in\mathbb{N}}$ in C_1.

Since C_1 is compact, there is a subsequence $(\sigma^{n_k})_{k\in\mathbb{N}}$ that converges to some $\sigma \in C_1$. The type σ is not null because we have

$$\| \sigma \| = \sigma(0) = \lim_k \sigma^{n_k}(0) = 1$$

Thus, there exists a nonzero type $\sigma \in C$ that is the limit of a p-approximating sequence.

If C' is the set of types τ belonging to C and the limit of a p-approximating sequence, then it is easy to see by a diagonal argument that C' is a closed conic class.

Since C is minimal, $C' = C$ and this proves Lemma III.4.7.

Lemma III.4.8. Let C be a closed minimal conic class. Then either C contains a nonzero c_0-type $\bar{\sigma}$, or there exists $p \in [1, +\infty[$ such that C contains a nonzero ℓ_p-type $\bar{\sigma}$.

Proof of Lemma III.4.8. Let Δ be a countable dense set in X. For α, β in $\mathbb{Q}^+$ and a in Δ we define a real function $\psi_{\alpha,\beta,a}$ on $\mathcal{S}_1(X)$ by

$$\forall \sigma \in \mathcal{S}_1(X), \qquad \psi_{\alpha,\beta,a}(\sigma) = \alpha\sigma * \beta\sigma \, (a)$$

In this way we get a countable family of functions of the first Baire class by Proposition III.1.5 and Lemma III.2.3.

If we consider the function ψ from $\mathcal{S}_1(X)$ into $\mathbb{R}^{\mathbb{Q} \times \mathbb{Q} \times \Delta}$ such that

$$\psi(\sigma) = (\psi_{\alpha,\beta,a}(\sigma))_{\alpha \in \mathbb{Q}, \beta \in \mathbb{Q}, a \in \Delta}$$

we still get a function of the first Baire class by [Ku], theorem $1'$, p. 382, because $\mathbb{Q} \times \mathbb{Q} \times \Delta$ is countable.

Set:

$$\mathcal{O} = \{\sigma \in C / 0 < \|\sigma\| < 1\}$$

Then $\mathcal{O}$ is an open set in C

Thus, by Theorem 1, p. 394 of [Ku], there exists a point of continuity $\bar{\sigma}$ of ψ in $\mathcal{O}$. That means that all functions $\psi_{\alpha,\beta,a}$ have a common point of continuity $\bar{\sigma}$ in $\mathcal{O}$.

Since $\bar{\sigma}$ is the limit of a p-approximating sequence $(\sigma^n)_{n \in \mathbb{N}}$ by Lemma III.4.7, the sequence $(\psi_{\alpha,\beta,a}(\sigma^n))_{n \in \mathbb{N}}$ converges to $\psi_{\alpha,\beta,a}(\bar{\sigma})$ for all $\alpha, \beta \in \mathbb{Q}^+$ and $a \in \Delta$.

If $p < +\infty$, this proves that

$$\alpha\bar{\sigma} * \beta\bar{\sigma}(a) = \lim_n (\alpha^p + \beta^p)^{1/p}\sigma^n(a) = (\alpha^p + \beta^p)^{1/p}\,\bar{\sigma}(a)$$

By an easy argument on density, we get that, for all $\alpha, \beta \in \mathbb{R}^+$ and $x \in X$,

$$\alpha\tilde{\sigma} * \beta\tilde{\sigma}(x) = (\alpha^p + \beta^p)^{1/p} \tilde{\sigma}(x)$$

This proves that $\tilde{\sigma}$ is an ℓ_p-type.

The case where $p = +\infty$ is analogous and we then get a c_0-type.

Proof of Theorem III.4.1. Lemma III.4.8 proves that there exists a nonzero c_0-type or ℓ_p-type for some $p \in [1, +\infty[$ in $\mathcal{T}(X)$. By Theorem II.4.2 and Remark III.4.3, it cannot be a c_0-type and, thus, there exists $p \in [1, +\infty[$ such that for all $\epsilon > 0$, X contains a subspace that is $(1 + \epsilon)$-isomorphic to ℓ_p.

Remark III.4.9. As a consequence of the proof of Theorem III.4.1, we get that any closed conic class on a stable space contains an ℓ_p-type for some $p \in [1, +\infty[$.

We also get the following:

Corollary III.4.10. If X is a stable Banach space that contains a subspace isomorphic to ℓ_p for some $p \in [1, +\infty[$, then for all $\epsilon > 0$ it contains a subspace that is $(1 + \epsilon)$-isomorphic to ℓ_p.

Proof. If $(x_n)_{n \in \mathbb{N}}$ is a good sequence in X, which is equivalent to the unit vector basis of ℓ_p, and if σ is the type defined by $(x_n)_{n \in \mathbb{N}}$, then in the conic class generated by $\sigma * -\sigma$, there is an ℓ_q-type for some q. It is easy to see by Theorem III.4.1 that for all $\epsilon > 0$ there exist disjoint blocks of $(x_n)_{n \in \mathbb{N}}$ which are $(1 + \epsilon)$-equivalent to the unit vector basis of ℓ_q. This implies that $q = p$ and proves Corollary III.4.10.

Remark. This result proves that the *distortion problem* (cf. the introduction) which is solved in the negative in general (as a very recent example of E. Odell and T. Schlumprecht shows) has a positive answer in the class of stable Banach spaces; namely, if stable space contains a subspace isomorphic to ℓ_p, it contains an almost isometric copy of this space.

5. OTHER PROPERTIES OF STABLE SPACES

The following property of stable spaces comes from [GL].

Theorem III.5.1.   Every stable Banach space X is weakly sequentially complete (w.s.c).

   We recall that the definition of w.s.c. was given in I.4.1.
   As a corollary of this theorem and of Corollary I.4.16, we get immediately

Corollary III.5.2.   Let X be a stable Banach space. Then either X contains a subspace, isomorphic to ℓ_1, or X is reflexive.

Proof of Theorem III.5.1.   Suppose that there exists a sequence $(x_n)_{n \in \mathbb{N}}$ in X, which is $\sigma(X,X^*)$-Cauchy and not $\sigma(X,X^*)$-convergent.

   By extraction, we can suppose that $(x_n)_{n \in \mathbb{N}}$ is basic (Theorem I.1.10) and that it is a good sequence (Proposition II.2.7). Let $(e_n)_{n \in \mathbb{N}}$ be the fundamental sequence of the spreading model, defined by $(x_n)_{n \in \mathbb{N}}$. By Remark III.3.2 there exists $K > 0$ such that $(e_n)_{n \in \mathbb{N}}$ is K-sign-unconditional.

   By Corollary II.3.2(c), $(e_n)_{n \in \mathbb{N}}$ is equivalent to the unit vector basis of ℓ_1.

   Consider for $n \in \mathbb{N}$

$$\begin{cases} y_n = x_{2n} - x_{2n-1} \\ f_n = e_{2n} - e_{2n-1} \end{cases}$$

Then $(y_n)_{n \in \mathbb{N}}$ is a good sequence in X which is $\sigma(X,X^*)$-convergent to 0; $(f_n)_{n \in \mathbb{N}}$ is the fundamental sequence of the spreading model defined by $(y_n)_{n \in \mathbb{N}}$ (cf. Lemma II.3.5).

   Since $(e_n)_{n \in \mathbb{N}}$ is equivalent to the unit vector basis of ℓ_1, $(f_n)_{n \in \mathbb{N}}$ is also equivalent to the unit vector basis of ℓ_1 and, on the other hand, because x is stable, it defines a symmetric type on X.

   We are going to prove that this situation is impossible, We need a lemma that extends the result of Theorem II.4.2:

Lemma III.5.3.   Let $(y_n)_{n \in \mathbb{N}}$ be a good sequence in X and $(f_n)_{n \in \mathbb{N}}$ the fundamental sequence of its spreading model. Suppose that $(y_n)_{n \in \mathbb{N}}$

defines a symmetric type σ on X and that there exists $p \in [1, +\infty[$ such that $(f_n)_{n \in \mathbb{N}}$ is K-equivalent to the unit vector basis of ℓ_p. Then there exist blocks $(z_n)_{n \in \mathbb{N}}$ on $(y_n)_{n \in \mathbb{N}}$, say $z_n = \sum_{i=p_n+1}^{p_{n+1}} \lambda_i y_i$, $\lambda_i \geq 0$ such that:

1. $(z_n)_{n \in \mathbb{N}}$ is equivalent to the unit vector basis of ℓ_p

2. $\dfrac{1}{K^p} \leq \sum_{i=p_n+1}^{p_{n+1}} \lambda_i^p \leq K^p$

Proof of Lemma III.5.3. Let $C = C(\sigma)$ be the closed conic class defined by σ. We divide the proof into two steps:

First step: We are going to prove that if $\tau \in C$, $\| \tau \| = 1$, then for all $(a_i)_{i=0,\ldots,k} \in \mathbb{R}^{k+1}$, we have

$$\frac{1}{K^2} \left(\sum_{i=0}^{k} | a_i |^p \right)^{1/p} \leq \| a_0 \tau * \cdots * a_k \tau \| \leq K^2 \left(\sum_{i=0}^{k} | a_i |^p \right)^{1/p}$$

Obviously, if τ is of the form $\alpha_0 \sigma * \cdots * \alpha_N \sigma$, it verifies this assumption with K instead of K^2.

Suppose that τ is limit of a normalized sequence $(\tau_n)_{n \in \mathbb{N}}$ of the form $\tau_n = \lambda_0^{(n)} \sigma * \cdots * \lambda_{p_n}^{(n)} \sigma$. Then, since $\| \tau_n \| = 1$, we have

$$\frac{1}{K} \left(\sum_{j=0}^{p_n} | \lambda_j^{(n)} |^p \right)^{1/p} \leq 1 \leq K \left(\sum_{j=0}^{p_n} | \lambda_j^{(n)} |^p \right)^{1/p}$$

Note that, since σ is symmetric, we can suppose that $\lambda_i^{(n)} \geq 0$ for all i and n.

Since the convolution product is separately continuous, we get

$$\| a_0 \tau * \cdots * a_k \tau \| = \lim_{n_0} \cdots \lim_{n_k} \| a_0 \tau_{n_0} * \cdots * a_k \tau_{n_k} \|$$

Then we can write

$$\frac{1}{K^2} \left(\sum_{i=0}^{k} | a_i |^p \right)^{1/p} \leq \frac{1}{K} \left(\sum_{i=0}^{k} \sum_{j=0}^{p_{ni}} | a_i \lambda_j^{(n)} |^p \right)^{1/p}$$

$$\leq \| a_0 \tau_{n_0} * \cdots * a_k \tau_{n_k} \|$$

$$\leq \left(\sum_{i=0}^{k} \sum_{j=0}^{p_{ni}} | a_i \lambda_j^{(n)} |^p \right)^{1/p} \leq K^2 \left(\sum_{i=0}^{k} | a_i |^p \right)^{1/p}$$

By taking the limits as above when $(n_0,\ldots,n_k)$ tends to $+\infty$, this proves the first step.

Second step: Let us show that the conclusion of Lemma III.5.3 holds. We know by Remark III.4.9 that there exists an ℓ_q-type τ_0 in C. The first step proves that $q = p$.

Let $(\tau_n)_{n \in \mathbb{N}}$ be a normalized sequence of types of the form $\tau_n = \lambda_0^{(n)}\sigma * \cdots * \lambda_{p_n}^{(n)}\sigma$, $\lambda_i^{(n)} \geq 0$, converging to τ_0.

For all $n \in \mathbb{N}$, define a sequence of blocks $(u_k^{(n)})_{k \in \mathbb{N}}$ on $(y_n)_{n \in \mathbb{N}}$ by

$$u_k^{(n)} = \sum_{i=0}^{p_n} \lambda_i^{(n)} y_{k(p_n+1)+i} \quad \text{for all } k \in \mathbb{N}$$

Then it is easy to see by a diagonal process that there exists a sequence $(u_{k\ell}^{(n_\ell)})_{\ell \in \mathbb{N}}$ that defines τ_0. If we set $z_\ell = u_{k\ell}^{(n_\ell)}$ for all $\ell \in \mathbb{N}$, the first step proves that $(z_\ell)_{\ell \in \mathbb{N}}$ verifies (1) and (2) of Lemma III.5.3.

We come back to the proof of Theorem III.5.2. Let $(z_n)_{n \in \mathbb{N}}$ be the blocks on $(y_n)_{n \in \mathbb{N}}$, given by Lemma III.5.3. Then, since $1/K \leq \sum_{i=p_n+1}^{p_{n+1}} \lambda_i \leq K$ and since $(y_n)_{n \in \mathbb{N}}$ is $\sigma(X,X^*)$-convergent to 0, $(z_n)_{n \in \mathbb{N}}$ is also $\sigma(X,X^*)$-convergent to 0. But $(z_n)_{n \in \mathbb{N}}$ is also equivalent to the unit vector basis of ℓ_1, which is a contradiction. Thus every $\sigma(X,X^*)$-Cauchy sequence in X is $\sigma(X,X^*)$-convergent and this proves that X is w.s.c.

NOTES AND REMARKS

1. The notion of stability of nonseparable spaces has been studied in [Ra1]. One of the definitions is that X is stable if all its separable subspaces are stable. This is analogous to Corollary I.6.3(2) for stable spaces. In [Ra1], the notion of superstability is also studied.

2. Recently, another proof of Theorem III.4.1 was found, using an ordinal index. See [Bu].

3. Weakly stable spaces have been introduced in [G] and [ANZ]. Every stable space is weakly stable and c_0 is weakly stable. These authors proved that every weakly stable space contains some ℓ_p, $1 \leq p < +\infty$, or c_0.

4. There are other known spaces that are stable: the trace classes C_p, $1 \leq p < +\infty$ [Ar]; quotients of ℓ_p, $1 \leq p < +\infty$ [C]; C_E for E stable [Ra3].

5. Banach spaces with strongly separable types, which are also a

generalization of stable spaces, were introduced and studied in [HM].

6. In [Gu2] and [Gu3] it is proved that there is no universal stable Banach space.

7. Other classes of Banach spaces are known to contain ℓ_p for some p—for instance, interpolation's spaces [BR] and [Lev] or spaces with a C^∞-smooth norm [De].

8. It was known that the *distortion problem* had a positive answer in subspaces of L_1 before the introduction of stable spaces; this is a result of Dacuhna-Castelle and Krivine [DCK2].

9. It is an open problem to characterize Banach spaces that have an equivalent stable norm.

EXERCISES

1. (a) Show that c_0 has no equivalent stable norm and deduce from this that no subspace of ℓ_1 is isomorphic to c_0.
   (b) Find an equivalent nonstable norm on ℓ_2.

2. Prove that if a stable space X has a spreading model isomorphic to ℓ_p, $1 \le p < +\infty$, then X contains a subspace isomorphic to ℓ_p.

3. Show that no spreading model on a stable space contains a subspace isomorphic to c_0.

4. Let X be a stable space and Y a finite-dimensional subspace of X. Show that the quotient space X/Y is stable.

5. Prove that if a stable space X contains a subspace isomorphic to ℓ_p, $1 \le p < +\infty$, then $\mathcal{T}(X)$ contains an ℓ_p-type.

6. Show that every spreading model over a stable space X is stable (see [GL2]).

7. Using a James' sequence (Chapter I.6), prove directly the following result: let X be a stable space; then X is reflexive if and only if it does not contain ℓ_1. (One can prove that the spreading model defined by the James' sequence is isomorphic to ℓ_1 and use Exercise 5.)

8. (See [Ra2].) Let $1 \le p < +\infty$. Prove that if X is stable, $\ell_p(X)$ is stable, directly as follows:
   (a) Let $(f_n)_{n \in \mathbb{N}}$, $(g_m)_{m \in \mathbb{N}} \subset \ell^p(X)$, $\| f_n \| \le 1$, $\| g_m \| \le 1$ for all m and n. Show that we can suppose that $f_n(i) = 0$ and $g_m(i) = 0$

for i large enough. Set

$$\alpha_j = \lim_{n,\mathcal{U}} \left(\sum_{i \geq j} \| f_n(i) \|_X^p \right), \qquad \beta_j = \lim_{m,\mathcal{V}} \left(\sum_{i \geq j} \| g_m(i) \|_X^p \right)$$

Prove that $(\alpha_j)_{j \in \mathbb{N}}$ and $(\beta_j)_{j \in \mathbb{N}}$ are convergent to some α and β, respectively.

Fix $\epsilon > 0$ and N_0 such that

$$j \geq N_0 \Rightarrow \begin{cases} \alpha \leq \alpha_j \leq \alpha + \epsilon^p \\ \beta \leq \beta_j \leq \beta + \epsilon^p \end{cases}$$

(b) Fix $n \in \mathbb{N}$ and j such that $f_n(i) = 0$ if $i \geq j$. Show that:

$$\lim_{m,\mathcal{V}} \| f_n + g_m \|_X^p = \lim_{m,\mathcal{V}} \left(\sum_{i \leq N_0} \| f_n(i) + g_m(i) \|_X^p \right.$$

$$\left. + \sum_{N_0 < i < j} \| f_n(i) + g_m(i) \|_X^p + \beta_j \right)$$

and prove that:

$$\lim_{m,\mathcal{V}} \left| \sum_{N_0 < i < j} \| f_n(i) + g_m(i) \|_X^p - \sum_{N_0 < i < j} \| f_n(i) \|_X^p \right| \leq 2^{p-1} p \epsilon$$

(c) Deduce from (b) that

$$\left| \lim_{m,\mathcal{V}} \| f_m + g_m \|_X^p - \lim_{m,\mathcal{V}} \sum_{i \leq N_0} \| f_n(i) - g_m(i) \|_X^p \right.$$

$$\left. - \sum_{i > N_0} \| f_n(i) \|_X^p - \beta_j \right| \leq 2^{p-1} p \epsilon$$

(d) Deduce from (b) that

$$\left| \lim_{n,\mathcal{U}} \lim_{n,\mathcal{V}} \| f_n + g_m \|_X^p - \lim_{n,\mathcal{U}} \lim_{m,\mathcal{V}} \sum_{i \leq N_0} \| f_n(i) - g_m(i) \|_X^p \right.$$

$$\left. - \alpha - \beta \right| \leq \epsilon^p + 2^{p-1} p \epsilon$$

(e) Prove that if X is stable, then $\ell_p(X)$ is stable.

9. Representation of functions $\mathbb{U}_i$ and $\mathbb{V}_i$ (see Theorem III.2.4) in $L_2(\Omega \mathcal{A}, P)$, $p \in 2\mathbb{N}$, or $1 \leq p < 2$ (see [Gu 4].

(a) $p = 2k$, $k \in \mathbb{N}^*$. Show that $f \in L_{2k} \Rightarrow f^i \in L_{2k/i}$.

Deduce from Newton's formula that for $f, g \in L_{2K}$:

$$\| f + g \|_{2k}^{2k} - \| f \|_{2k}^{2k} - \| g \|_{2k}^{2k} = \sum_{i=1}^{2k-1} C_{2k}^i \langle f^i, g^{2k-1} \rangle$$

where $\langle \, , \, \rangle$ is the duality product in $(L^{2k/i}, L^{2k/2k-i})$. What are functions $\mathbb{U}_i$ and $\mathbb{V}_i$ and reflexive spaces $\mathcal{E}_i$ and $\mathcal{E}_i^*$ in this case?

(b) $1 \leq p < 2$. Show that there exists $0 < K_p < +\infty$ such that

$$\forall x \in \mathbb{R}, \qquad |x|^p = K_p \int_0^{+\infty} (1 - \cos tx) \frac{dx}{x^{p+1}}$$

Developing $\cos(x + y)$, show that for $f, g \in L_p$:

$$\| f - g \|^p - \| f \|^p - \| g \|^p = K_p \, [-\langle Uf, Ug \rangle - \langle Vf, Vg \rangle]$$

where

$$Uf = 1 - \cos tf \in L_2 \left(\Omega \times [0, +\infty[, \, dP \otimes \frac{dt}{t^{p+1}} \right)$$

$$Vf = \sin tf \in L_2 \left(\Omega \times [0, +\infty[, \, dP \otimes \frac{dt}{t^{p+1}} \right)$$

(Prove that the integrals converge.) $\langle \, , \, \rangle$ denotes the inner product in $L_2(\Omega \times [0, +\infty[)$ above.

(c) Is (a) true for $p \in \mathbb{N}$, p odd? Is (b) true for $p > 2$?

IV

Subspaces of L_p-Spaces, $1 \leq p < +\infty$

In Chapter III we proved that L_p-spaces are stable (Corollary III.2.7). Thus, every subspace of L_p-spaces contains ℓ_q for some $q \in [1, +\infty[$ and does not contain c_0. In this chapter, we are going to give more precise conditions on the set of q's such that ℓ_q embeds in L_p or in a given subspace of L_p.

The aim of the first three sections of this chapter is to present properties of L_p-spaces and their ultrapowers, which will be necessary to compute these values of q.

In the two last sections, we will give the final computation of the values of q.

1. REDUCTION TO THE CASE OF ℓ_p and $L_p[0,1]$

In this section we show that it is sufficient, up to an isometry, to work with $L_p[0,1]$ and ℓ_p instead of general L_p-spaces. To avoid trivial cases, we will always suppose that L_p-spaces are infinite dimensional.

The main results of this part can be found in P. R. Halmos's book

[Ha] for the measure theory part and in H. E. Lacey's book [L] for the Banach space theory part.

Definition and Notation

1. $L_p = L_p(\Omega,\Sigma,\mu)$ will denote the usual L_p-space on the measure space (Ω,Σ,μ) and $L_p[0,1]$ will denote $L_p([0,1], \mathcal{B}, \lambda)$ where $\mathcal{B}$ is the Borel σ-field on $[0,1]$ and λ the Lebesgue measure. We will use dt instead of $d\lambda(t)$ when there is no ambiguity.
2. If E are F are two Banach spaces, and $p \in [1, +\infty[$, then we define the Banach space $E \oplus_p F$ by

$$E \oplus_p F = \{(x,y) \in E \times F / \| (x,y) \| = (\| x \|_E^p + \| y \|_F^p)^{1/p}\}$$

First we are going to work in the separable case.

Theorem IV.1.1.   Let $1 \le p < +\infty$ and let $L_p = L_p(\Omega,\Sigma,\mu)$ be a separable L_p-space. Then there exists a finite or countable set I such that L_p is isometric to $L_p[0,1] \oplus_p \ell_p(I)$.

Proof.   We need four steps.

Step 1: Let us prove that μ is σ-finite. Let $(f_n)_{n\in\mathbb{N}}$ be a countable dense set in $L_p(\Omega,\Sigma,\mu)$. For all $n \in \mathbb{N}$, there exists a simple function g_n such that $\| f_n - g_n \|_p \le 1/2^n$. Let us state $g_n = \sum_{i\in I_n} a_i^{(n)} 1_{A_i^{(n)}}$ where I_n is a finite set, $a_i^{(n)}$ are scalars, and $A_i^{(n)}$ belong to Σ, are disjoint, and have finite measure. Then the sequence $(g_n)_{n\in\mathbb{N}}$ is also a dense set in L_p. Let Σ' be the the σ-field generated by the countable set

$$\{A_i^{(n)}, i \in I_n, n \in \mathbb{N}\}$$

Then it is clear that $L_p(\Omega,\Sigma,\mu) = L_p(\Omega,\Sigma',\mu)$ and μ is σ-finite.

Step 2: By a change of density, we are going to prove that we can work with $L_p(\Omega,\Sigma',\mu')$ where μ' is a probability measure.

Let $f \in L^p(\Omega,\Sigma',\mu)$ such that $f > 0$ and $\int_\Omega f^p \, d\mu = 1$. The existence of such an f is a consequence of the fact that μ is σ-finite: indeed, let Σ' be generated by a countable infinite family $\{A_n, n \in \mathbb{N}\}$ with $A_n \cap A_m = \varnothing$ if $n \ne m$ and $\mu(A_n) \ne 0$ for all $n \in \mathbb{N}$. This family exists as long as $L_p(\Omega,\Sigma',\mu)$ is not finite dimensional, which we already supposed.

Define:

$$f = \sum_{n=0}^{\infty} \frac{1}{2^{n+1}} \frac{1_{A_n}}{\mu(A_n)}$$

This function f is suitable. Next, define μ' by $\mu' = f^p \mu$. Then μ' is a probability measure and the operator

$$T : L^p(\Omega, \Sigma', \mu') \rightarrow L^p(\Omega, \Sigma', \mu')$$

$$g \mapsto gf^{-1}$$

is obviously an isometry.

Step 3: Let us decompse μ' into a nonatomic part μ_1 and a sum of atomic measures.

The σ-field Σ' may contain atoms. Let us call them $(A_i)_{i \in I}$. Since μ' is a probability measure, the sum $\sum_{i \in I} \mu'(A_i)$ is finite. Thus I is countable and we can suppose without loss of generality that $I = \mathbb{N}$. The countable union $\cup_{n \in \mathbb{N}} A_n$ is measurable and so is $\Omega_1 = \Omega \backslash \cup_{n \in \mathbb{N}} A_n$. Denote by μ_1 and Σ_1 the restrictions of μ' and Σ' to Ω_1. Then μ_1 is nonatomic on Σ_1.

Every function f in $L_p(\Omega, \Sigma', \mu')$ can be written as a couple $(f_1, (a_n)_{n \in \mathbb{N}})$ where $f_1 = f_{|\Omega_1}$ and $a_n = f_{|A_n} \mu(A_n)$. The norm of f in $L_p(\Omega, \Sigma', \mu')$ is then

$$\| f \|_p^p = \int_{\Omega_1} | f_1 |^p \, d\mu_1 + \sum_{n=0}^{\infty} | f_{|A_n} |^p \mu(A_n)$$

$$= \int_{\Omega_1} | f_1 |^p \, d\mu_1 + \sum_{n=0}^{\infty} | a_n |^p$$

This proves that we can write

$$L_p(\Omega, \Sigma', \mu') = L_p(\Omega_1, \Sigma_1, \mu_1) \underset{p}{\oplus} \ell_p$$

Step 4: It remains to prove that $L_p(\Omega_1, \Sigma_1, \mu_1)$ is null or isometric to $L_p[0,1]$.

First of all, note that if $L_p(\Omega_1, \Sigma_1, \mu_1)$ is not null, that is, if μ_1 is not the null measure, then by another change of density (see step 2) again we can suppose that μ_1 is a nonatomic probability measure.

We need a definition and two lemmas:

Definition IV.1.2

1. For any nonatomic probability measure v on $(\Omega,\mathcal{A})$ and $p \in [1, +\infty[$, we define a *distance* on $\mathcal{A}$ by

$$\forall A,A' \in \mathcal{A}, \qquad d(A,A') = v(A \Delta A') = \| 1_A - 1_{A'} \|_p^p$$

2. A family $(A_i)_{i \in I}$ of elements of $\mathcal{A}$ is said to be *dense* in $\mathcal{A}$ if for all $\epsilon > 0$ and all A in $\mathcal{A}$, there exists $i \in I$ such that

$$d(A,A_i) \leq \epsilon$$

3. If $(\mathcal{P}_n)_{n \in \mathbb{N}}$ is a sequence of finite and $\mathcal{A}$-measurable partitions of Ω, we will say that

   (a) $(\mathcal{P}_n)_{n \in \mathbb{N}}$ is *decreasing* if for all P in $\mathcal{P}_n$ there exists Q in $\mathcal{P}_{n-1}$ such that $P \subset Q$.

   (b) $(\mathcal{P}_n)_{n \in \mathbb{N}}$ is *dense* if the set of all finite unions of elements of $(\mathcal{P}_n)_{n \in \mathbb{N}}$ is dense in $\mathcal{A}$ for d.

4. $v(\mathcal{P}_n) = \text{Sup}\{v(P)/P \in \mathcal{P}_n\}$.

Remark IV.1.3

1. It is clear by Lebesgue's theorem that $(\mathcal{A},d)$ is a complete metric space.
2. It is easy to see that $L_p(\Omega,\mathcal{A},v)$ is separable if and only if $(\mathcal{A},d)$ is a separable metric space.

With this definition, we can state the two lemmas that we need to prove step 4:

Lemma IV.1.4. If $(\mathcal{P}_n)_{n \in \mathbb{N}}$ is a decreasing and dense sequence of finite and Σ_1-measurable partitions of Ω_1, then $\lim_n \mu_1(\mathcal{P}_n) = 0$.

Proof. $(\mu_1(\mathcal{P}_n))_{n \in \mathbb{N}}$ is a decreasing sequence of positive numbers. Let δ be its limit and suppose that $\delta > 0$.

It is clear that there exists a decreasing sequence $P_1 \supset P_2 \supset \cdots \supset P_k \supset \cdots$ such that $P_k \in \mathcal{P}_k$ and $\mu_1(P_k) \geq \delta$ for all $k \in \mathbb{N}$. Set $P = \cap_{k \in \mathbb{N}} P_k$. Then $\mu_1(P) \geq \delta$. But since μ_1 has no atom, there exists $P_0 \subset P$ with $0 < \mu_1(P_0) < \mu_1(P)$. We observe that P_0 is contained in P_k for all k and is disjoint from all the other elements of $\mathcal{P}_k$.

It follows that if ϵ is smaller than $\mu_1(P_0)$ and $\mu_1(P \backslash P_0)$, then no element of Σ_1, which is a union of elements of $(\mathcal{P}_n)_{n\in\mathbb{N}}$ can have a distance less that ϵ from P_0. This contradicts the density of $(\mathcal{P}_n)_{n\in\mathbb{N}}$ and thus $\delta = 0$.

Lemma IV.1.5. Let $(\mathcal{Q}_n)_{n\in\mathbb{N}}$ be a sequence of finite partitions into subintervals of $[0,1]$ such that $\lim_n \lambda(\mathcal{Q}_n) = 0$. Then $(\mathcal{Q}_n)_{n\in\mathbb{N}}$ is dense in the Borel σ-field $\mathcal{B}$.

Proof. To every $\epsilon > 0$, there corresponds $n \in \mathbb{N}$ such that $\lambda(\mathcal{Q}_n) \leq \epsilon/2$. Let I be an interval of $[0,1]$ and let I_1 be the unique element of $\mathcal{Q}_n$ which contains the left end point of I. If I_1 does not contain I, let I_2 be the interval of $\mathcal{Q}_n$ which is adjacent to I_1 to the right. Reiterating this procedure a finite number of times, we get an interval $I_k \in \mathcal{Q}_n$ which contains the right end point of I. Then

$$\int_0^1 |1_I - 1_{\cup_{i=1}^k I_i}|^p \, d\lambda = d\left(I, \bigcup_{i=1}^k I_i\right) \leq \epsilon$$

Since the class of all finite unions of intervals is dense in $\mathcal{B}$, the proof is complete.

Let us prove step 4: By Remark IV.1.3.2, we can find a dense set $(A_n)_{n\in\mathbb{N}}$ in Σ_1. For all $n \in \mathbb{N}$, define a partition $\mathcal{P}_n$ of Ω_1 by

$$\mathcal{P}_n = \{\bigcap_{i\in I} A_i \cap {}^c A_j / I \cap J = \varnothing, I \cup J = \{0, \ldots, n\}\}$$

Then $(\mathcal{P}_n)_{n\in\mathbb{N}}$ is a decreasing and dense sequence of finite partitions of Ω_1. Lemma IV.1.4 implies that $\lim_n \mu_1(\mathcal{P}_n) = 0$.

We define a transform T by the following procedure: For all $P \in \mathcal{P}_0$, we define $T(P)$ such that

$$\begin{cases} T(P) \text{ is an interval in } [0,1] \\ \mu_1(P) = \lambda(T(P)) \\ Q_0 = \{T(P), P \in \mathcal{P}_0\} \text{ is a partition of } [0,1] \end{cases}$$

For each $P \in \mathcal{P}_0$, $P \cap \mathcal{P}_1 = \{P \cap P_1, P_1 \in \mathcal{P}_1\}$ is a partition of P. So for all $P \cap P_1 \in P \cap \mathcal{P}_1$, we define $T(P \cap P_1)$ such that

$$\begin{cases} T(P \cap P_1) \text{ is an interval in } T(P) \\ \mu_1(P \cap P_1) = \lambda(T(P \cap P_1)) \\ Q_1 = \{T(P \cap P_1), P_1 \in \mathcal{P}_1\} \text{ is a partition of } T(P) \end{cases}$$

Continuing this construction, we define a sequence of finite partitions $(\mathfrak{Q}_n)_{n\in\mathbb{N}}$ of $[0,1]$ such that $\lim_n \lambda(\mathfrak{Q}_n) = 0$. Lemma IV.1.5 implies that $(\mathfrak{Q}_n)_{n\in\mathbb{N}}$ is dense in $\mathfrak{B}$.

It can easily be seen that T can be extended to all $A \in \Sigma_1$, then to all simple functions in $L_p(\Omega_1,\Sigma_1,\mu_1)$, and thus to all $L_p(\Omega_1,\Sigma_1,\mu_1)$. We obtain in this way an isometry from $L_p(\Omega_1,\Sigma_1,\mu_1)$ onto $L_p([0,1], \mathfrak{B}, \lambda)$. This finishes the proof of Theorem IV.1.1.

Corollary IV.1.6. For $1 \le p < +\infty$, every separable L_p-space is isometric to a subspace of $L_p[0,1]$.

Proof. Let $L_p = L_p(\Omega,\Sigma,\mu)$ be a separable L_p-space. By Theorem IV.1.1 we can write $L_p = L_p[0,1] \oplus \ell_p(I)$ for some finite or countable set I.

Let us show that $\ell_p(I)$ *is isometric to a subspace of* $L_p[0,1]$: since I is finite or countable, there exists a family $(I_i)_{i\in I}$ of mutually disjoint intervals of $[0,1]$. Then the closed linear span of $(1_{I_i}/\lambda(I_i)^{1/p})_{i\in I}$ is isometric to $\ell_p(I)$. Thus $L_p[0,1] \oplus_p \ell_p(I)$ embeds isometrically in $L_p[0,1] \oplus_p L_p[0,1]$, which is isometric to $L_p([0,1] \times [0,1], \lambda \otimes \lambda)$, which itself is isometric to $L_p([0,1])$ by Theorem IV.1.1.

This proves Corollary IV.1.6.

We are now going to study nonseparable L_p-spaces. A complete representation of nonseparable L_p-spaces, similar to the representation of separable L_p-spaces, can be found in H. E. Lacey's book [L]. However, we will not need the full strength of this result here and we are just going to give the following property, which is much easier and will be enough for this part:

Theorem IV.1.7. Let $1 \le p < +\infty$ and $L_p = L_p(\Omega,\Sigma,\mu)$. Then every separable subspace of L_p is isometric to a subspace of $L_p[0,1]$.

Proof. Let X be a separable subspace of L_p and $(f_n)_{n\in\mathbb{N}}$ a dense subset of X. As in step 1 of Theorem IV.1.1, we can suppose that the f_n's are simple functions, namely $f_n = \sum_{i\in I_n} a_i^n A_i^n$, with $a_i^n \in \mathbb{R}$, $|I_n| < +\infty$, and $A_i^n \in \Sigma$. Let us call Σ' the σ-field generated by the sets A_i^n, for $i \in I_n$ and $n \in \mathbb{N}$. Then X is a subspace of $L_p(\Omega,\Sigma',\mu)$ which is separable. By Corollary IV.1.6, $L_p(\Omega,\Sigma',\mu)$ is isometric to a subspace of L_p [0,1] and this proves Theorem IV.1.7.

2. BASIC LEMMAS IN $L_p[0,1]$, $1 \le p < +\infty$

The first idea in this section is p-equi-integrability. In the case $p = 1$, it is related to weak compactness. These results can be found in [LT], [Be3], and [N] for the probabilistic part.

Definition IV.2.1. Let $p \in [0,+\infty[$ and let F be a subset of $L_p[0,1]$. F is said to be *p-equi-integrable* if $\text{Sup}_{f \in F}\{\int_{\{|f|>a\}} |f(t)|^p \, dt\}$ tends to 0 when $a \to +\infty$.

Let us give some easy examples of p-equi-integrable sets:

Examples

1. If F is a finite subset of $L_p[0,1]$, then F is p-equi-integrable.
2. A finite union of p-equi-integrable sets is p-equi-integrable.
3. If $(X_n)_{n \in \mathbb{N}}$ is a sequence of i.i.d. (independent and identically distributed) random variables in $L_p[0,1]$, $(X_n)_{n \in \mathbb{N}}$ is p-equi-integrable.
4. Set $f_n = n^{1/p} 1_{[0,1/n]}$. Then $F = (f_n)_{n \in \mathbb{N}}$ is not p-equi-integrable.

The next result also gives an example of p-equi-integrable sets:

Proposition IV.2.2. Let $1 \le p < +\infty$ and suppose that F is a subset of $L_p[0,1]$. If there exists $g \in L_p[0,1]$ such that for all f in F, $|f| \le g$, then F is p-equi-integrable.

Proof. We can write

$$\int_{|f|>a} |f(t)|^p \, dt \le \int_{|g|>a} |g(t)|^p \, dt = \|g\|_p^p - \int_{|g|\le a} |g(t)|^p \, dt$$

By Lebesgue's theorem, $\int_{|g|\le a} |g(t)|^p \, dt$ converges to $\|g\|_p^p$ when $a \to +\infty$ and this proves this proposition.

The next proposition gives a very useful characterization of p-equi-integrability:

Proposition IV.2.3. For $1 \leq p < +\infty$, let F be a subset of $L_p[0,1]$. Then F is p-equi-integrable if and only if the following two properties are verified:

(a) For every $\epsilon > 0$, there exists $\delta > 0$ such that for all $A \in \mathcal{B}$ with $\lambda(A) < \delta$, then $\forall f \in F$, $\int_A |f|^p \leq \epsilon$.

(b) There exists $M \in \mathbb{R}^+$ such that $\mathrm{Sup}_{f \in F} \| f \|_p \leq M$.

Proof. Suppose that F is p-equi-integrable. Then, for $f \in F$, $A \in \mathcal{B}$, and $a \in \mathbb{R}^+$, we can write

$$\int_A | f(t)|^p \, dt = \int_{\{|f| \leq a\} \cap A} | f(t)|^p \, dt + \int_{\{|f| > a\} \cap A} | f(t)|^p \, dt$$

$$\leq a^p \, \lambda(A) + \int_{\{|f| > a\}} | f(t)|^p \, dt$$

Fix $\epsilon > 0$. By definition, there exists $a_0 \in \mathbb{R}^+$ such that for all f in F,

$$\int_{\{|f| > a_0\}} | f(t)|^p \, dt \leq \frac{\epsilon}{2}$$

Choose $\delta > 0$ such that $a_0^p \, \delta \leq \epsilon/2$. For this choice, we get, for all $A \in \mathcal{B}$, with $\lambda(A) < \delta$,

$$\int_A | f(t)|^p \, dt \leq \epsilon/2 + \epsilon/2 = \epsilon$$

$$\int | f(t)|^p \, dt \leq a_0^p + \epsilon/2$$

which proves (a) and (b).

On the other hand, suppose that (a) and (b) are true. Recall *Tchebychev's inequality*:

For all f in $L_p[0,1]$, we have $\lambda\{| f | > a\} \leq \dfrac{\| f \|_p^p}{a^p}$

Then, for $f \in F$, $\lambda\{| f | > a\} \leq M^p/a^p$, which tends to 0 when $a \to +\infty$. Fix $\epsilon > 0$. By (a) there is $\delta > 0$ such that, if $\lambda(A) \leq \delta$, then for all $f \in F$, $\int_A | f(t)|^p \, dt \leq \epsilon$.

Choose a_0 such that $M < a_0^p \, \delta$. Then, for all $f \in F$, $\lambda\{| f | > a\} \leq \delta$ and thus

$$\int_{\{|f| > a\}} | f(t)|^p \, dt \leq \epsilon$$

This proves that F is p-equi-integrable.

When $p \in]1, +\infty[$, $L_p[0,1]$ is reflexive and bounded sets are relatively $\sigma(L_p, L_{p'})$ compact.

In the particular case where $p = 1$, since $L_1[0,1]$ is not reflexive, bounded sets are not necessarily relatively $\sigma(L_1, L_\infty)$-compact.

We are going to characterize $\sigma(L_1, L_\infty)$-compact sets of $L_1[0,1]$ in terms of 1-equi-integrability.

Theorem IV.2.4. Let $(f_n)_{n \in \mathbb{N}}$ be a sequence in $L_1[0,1]$.

1. If $\lim_n \int_A f_n(t) \, dt$ exists for all $A \in \mathcal{B}$, then $(f_n)_{n \in \mathbb{N}}$ is 1-equi-integrable.
2. $\lim_n \int_A f_n(t) \, dt$ exists for all $A \in \mathcal{B}$ if and only if $(f_n)_{n \in \mathbb{N}}$ is $\sigma(L_1, L_\infty)$-convergent.

Proof. For any $f \in L_1[0,1]$, define a map φ by

$$\varphi \colon \mathcal{B} \to \mathbb{R}$$
$$A \mapsto \int_A f(t) \, dt$$

Lemma IV.2.5. φ is uniformly continuous on the metric space $(\mathcal{B}, d)$ (see Definition IV.1.2).

Proof of Lemma IV.2.5. For $A, A' \in \mathcal{B}$, we have

$$\left| \int_A f(t) \, dt - \int_{A'} f(t) \, dt \right| = \left| \int_{A \backslash A'} f(t) \, dt - \int_{A' \backslash A} f(t) \, dt \right|$$
$$\leq \int_{A \Delta A'} |f(t)| \, dt$$

Since $f \in L_1[0,1]$, for all $\epsilon > 0$ there exists $\delta > 0$ such that

$$\lambda(B) \leq \delta \Rightarrow \int_B |f(t)| \, dt \leq \epsilon$$

Thus if $\lambda(A \, \Delta A') = d(A, A') \leq \delta$, then:

$$\left| \int_A f(t) \, dt - \int_{A'} f(t) \, dt \right| \leq \int_{A \Delta A'} |f(t)| \, dt \leq \epsilon$$

We come back to the proof of Theorem IV.2.4:

Proof of (1): If $(f_n)_{n \in \mathbb{N}}$ is a sequence in $L_1[0,1]$ such that $\lim_n \int_A f_n(t) \, dt$

exists for all $A \in \mathcal{B}$, $\epsilon > 0$, and $N \in \mathbb{N}$, we define F_N by

$$F_N = \left\{ A \in \mathcal{B} / \forall m,n \geq N, \left| \int_A (f_m(t) - f_n(t)) \, dt \right| \leq \epsilon \right\}$$

By Lemma IV.2.5, F_N is closed in $(\mathcal{B},d)$ and by hypothesis $\mathcal{B} = \cup_{N \in \mathbb{N}} F_N$. Thus since $(\mathcal{B},d)$ is a complete metric space (see Remark IV.1.3), by Baire's theorem there exists $N_0 \in \mathbb{N}$ such that int $F_{N_0} \neq \varnothing$. That means that there exist $A_0 \in \mathcal{B}$ and $\delta > 0$ such that, for all $A \in \mathcal{B}$,

$$\lambda(A \Delta A_0) < \delta \Rightarrow \forall m,n \geq N_0, \quad \left| \int_A (f_m(t) - f_n(t)) \, dt \right| \leq \epsilon$$

Let $B \in \mathcal{B}$. We can write

$$\int_B (f_m(t) - f_n(t)) \, dt$$

$$= \int_{A_0 \cup B} (f_m(t) - f_n(t)) \, dt - \int_{A_0 \backslash B} (f_m(t) - f_n(t)) \, dt$$

But $(A_0 \cup B) \, \Delta \, A_0 \subset B$ and $(A_0 \backslash B) \, \Delta \, A_0 \subset B$.
Thus, if $\lambda(B) < \delta$, we get

$$\int_B (f_m(t) - f_n(t)) \, dt \leq 2\epsilon \quad \text{for all } n,m \geq N_0$$

Applying this to $B \cap \{f_n \geq f_m\}$ and $B \cap \{f_m \geq f_n\}$, we also get that

$$\int_B | f_m(t) - f_n(t) | \, dt \leq 4\epsilon \quad \text{for all } n,m \geq N_0$$

On the other hand, since f_{N_0} belongs to $L_1[0,1]$, there exists $\delta' > 0$ such that

$$\lambda(B) < \delta' \Rightarrow \int_B | f_{N_0}(t)| \, dt \leq \epsilon$$

Taking $\delta_0 = \text{Inf} \, (\delta,\delta')$, if $\lambda(B) \leq \delta_0$, we get

$$n \geq N_0 \Rightarrow \int_B | f_n(t) | \, dt \leq \int_B | f_n(t) - f_{N_0}(t) | \, dt + \int_B | f_{N_0}(t) | \, dt$$

And then: $\qquad\qquad\qquad \leq 4\epsilon + \epsilon = 5\epsilon$

$$n \geq N_0 \Rightarrow \int | f_n(t) | \, dt \leq 5\epsilon \left(\left[\frac{1}{\delta_0} \right] + 1 \right)$$

Thus $(f_n)_{n \geq N_0}$ verifies (a) and (b) of Proposition IV.2.3, so it is 1-equi-integrable and $(f_n)_{n \in \mathbb{N}}$, as a union of two 1-equi-integrable sets, is also 1-equi-integrable.

Proof of (2): Suppose that $\lim_n \int_A f_n(t)\, dt$ exists for all $A \in \mathcal{B}$. Define a map μ from $\mathcal{B}$ to $\mathbb{R}$ by $\forall A \in \mathcal{B}$ $\mu(A) = \lim_n \int_A f_n(t)\, dt$. Let us prove that μ is σ-additive:

If $A = \bigcup_{k=0}^{\infty} A_k$ with $A_k \cap A_l = \varnothing$ if $k \neq l$, then $\lambda(A) = \sum_{k=0}^{\infty} \lambda(A_k)$. Hence for $\delta > 0$ there exists $N_0 \in \mathbb{N}$ such that

$$P,Q \geq N_0 \Rightarrow \sum_{k=P}^{Q} \lambda(A_k) = \lambda \left(\bigcup_{k=P}^{Q} A_k \right) \leq \delta$$

Since we can write

$$\left| \sum_{k=P}^{Q} \int_{A_k} f_n(t)\, dt \right| \leq \int_{\bigcup_{k=P}^{Q} A_k} |f_n(t)|\, dt$$

the 1-equi-integrability of $(f_n)_{n \in \mathbb{N}}$ implies that the series $\sum_{k=0}^{\infty} \int_{A_k} f_n(t)\, dt$ is uniformly convergent for $n \in \mathbb{N}$. So we get

$$\mu(A) = \lim_n \int_A f_n(t)\, dt = \lim_n \sum_{k=0}^{\infty} \int_{A_k} f_n(t)\, dt$$

$$= \sum_{k=0}^{\infty} \lim_n \int_{A_k} f_n(t)\, dt = \sum_{k=0}^{\infty} \mu(A_k)$$

Thus μ is a measure on $([0,1], \mathcal{B})$.

Since $(f_n)_{n \in \mathbb{N}}$ is 1-equi-integrable, we get that for all $\epsilon > 0$, there exists $\delta > 0$ such that $\lambda(A) < \delta \Rightarrow \int_A |f_n(t)|\, dt \leq \epsilon$. This implies also that $\mu(A) \leq \epsilon$. Thus μ is absolutely continuous with respect to λ.

By Radon–Nikodym's theorem (see, for example, [Ha], p. 128), there exists f in $L_1[0,1]$ such that for all A in $\mathcal{B}$, $\mu(A) = \int_A f(t)\, dt$. It is clear that if g is a simple function, then, by definition, the sequence $(\int f_n(t)g(t)\, dt)_{n \in \mathbb{N}}$ converges to $\int f(t)g(t)\, dt$. Since simple functions are dense in $L_\infty[0,1]$, if $g \in L_\infty[0,1]$ there exists a sequence $(g_k)_{k \in \mathbb{N}}$ of simple functions such that

$$\| g - g_k \|_\infty \xrightarrow[k \to +\infty]{} 0$$

Hence:

$$\left| \int f_n(t)g(t)\, dt - \int f(t)g(t)\, dt \right| \leq \left| \int f_n(t)g(t)\, dt - \int f_n(t)g_k(t)\, dt \right|$$

$$+ \left| \int f_n(t)g_k(t)\, dt - \int f(t)g_k(t)\, dt \right|$$

$$+ \left| \int f(t)g_k(t)\, dt - \int f(t)g(t)\, dt \right|$$

$$\leq \| f_n \|_1 \| g_k - g \|_\infty$$

$$+ \left| \int (f_n(t) - f(t))g_k(t)\, dt \right|$$

$$+ \| f \|_1 \| g_k - g \|_\infty$$

By (1), there exists M such that $\| f_n \|_1 \leq M$ and $\| f \|_1 \leq M$. Fix $\epsilon > 0$ and choose k_0 such that $2M \| g_{k_0} - g \|_\infty \leq \epsilon/2$. Choose N_0 such that

$$n \geq N_0 \Rightarrow | \int (f_n(t) - f(t))g_{k_0}(t)\, dt| \leq \epsilon/2$$

Then the preceding property implies that

$$n \geq N_0 \Rightarrow \left| \int f_n(t)g(t)\, dt - \int f(t)g(t)\, dt \right| \leq \epsilon$$

since the converse property is obvious, this proves (2).

Corollary IV.2.6. $L_1[0,1]$ is w.s.c.

Recall that w.s.c. was defined in Definition I.4.1.

Proof.   Since L_1 is stable, Corollary III.2.7, this property is a consequence of Theorem III.5.1. However, let us give a direct and easy proof of it which is a direct application of the preceding results:

If $(f_n)_{n \in \mathbb{N}}$ is a $\sigma(L_1, L_\infty)$-Cauchy sequence in L_1, then for all $A \in \mathcal{B}$, the real sequence $(\int_A f_n(t)\, dt)_{n \in \mathbb{N}}$ is convergent. By Theorem IV.2.4.2, $(f_n)_{n \in \mathbb{N}}$ is $\sigma(L_1, L_\infty)$-convergent.

Corollary IV.2.7.   Let F be a subset of $L_1[0,1]$. F is relatively $\sigma(L_1, L_\infty)$-compact if and only if F is 1-equi-integrable.

Proof.   Suppose that F is not 1-equi-integrable. Then there exist $\epsilon_0 > 0$ and a sequence $(f_n)_{n \in \mathbb{N}}$ in F such that

$$\int_{\{|f_n| > n\}} | f_n(t)|\, dt \geq \epsilon_0$$

This sequence has no 1-equi-integrable subsequence. Thus, by Theorem IV.2.4 it has no $\sigma(L_1,L_\infty)$-convergent subsequence. This proves by Eberlein–Šmŭlian's theorem (see [DS] Vol I) that F is not relatively $\sigma(L_1,L_\infty)$-compact.

Suppose now that F is 1-equi-integrable. Then F is a bounded subset of $L_1[0,1]$ and identifying $L_1[0,1]$ with a subspace of $(L_1[0,1])^{**} = (L_\infty[0,1])^*$, for all sequences $(f_n)_{n\in\mathbb{N}}$ in F we can extract a subsequence $(f_{n_k})_{k\in\mathbb{N}}$ which converges for $\sigma(L_\infty^*,L_\infty)$ to some $\mu \in L_\infty^*$. In particular, we have

$$\forall A \in \mathcal{B}, \qquad \mu(A) = \lim_k \int_A f_{n_k}(t)\, dt$$

Applying Theorem IV.2.4.2, we see that $(f_{n_k})_{k\in\mathbb{N}}$ is $\sigma(L_1,L_\infty)$-convergent. So by Eberlein–Šmŭlian's theorem, F is relatively compact.

We are now going to present the so-called *subsequence splitting lemma*. It is a technical result which explains the behavior of bounded sequences in $L_p[0,1]$. The proof of this result was first given by H. P. Rosenthal in a course that he gave in 1979 at the University Paris 6. It is a refinement of arguments buried in the classical paper of M. Kadec and A. Pelczyinski [KP1].

We need to recall some well-known notation:

Notation. Let f, g, h belong to $L_p[0,1]$; the notation $|\,g\,| \wedge |\,h\,| = 0$ means that the function g and h have disjoint supports on $[0,1]$. Supp f denotes the support of the function f on $[0,1]$, that is, the set where f is nonzero.

Theorem IV.2.8. Let $1 \leq p < +\infty$ and $(f_n)_{n\in\mathbb{N}}$ be a bounded sequence in $L_p[0,1]$. There exist a subsequence $(f_{n_k})_{k\in\mathbb{N}}$ of $(f_n)_{k\in\mathbb{N}}$ and two sequences $(g_k)_{k\in\mathbb{N}}$ and $(h_k)_{k\in\mathbb{N}}$ such that

(a) $\forall k \in \mathbb{N}$, $f_{n_k} = g_k + h_k$
(b) $|\,g_k\,| \wedge |\,h_k\,| = 0$
(c) $(g_k)_{k\in\mathbb{N}}$ is p-equi-integrable
(d) $\lambda(\text{Supp } h_k) \xrightarrow[k\to+\infty]{} 0$

The proof of Theorem IV.2.8 is an easy consequence of the following lemma:

Lemma IV.2.9. Let $(f_n)_{n\in\mathbb{N}}$ be a bounded sequence in $L_1[0,1]$. There exist a subsequence $(f_{n_k})_{k\in\mathbb{N}}$ of $(f_n)_{n\in\mathbb{N}}$ and a sequence $(A_k)_{k\in\mathbb{N}}$ in $\mathscr{B}$ such that

$$\left[\begin{array}{l} \lambda(A_k) \xrightarrow[k\to+\infty]{} 0 \\ (f_{n_k}1_{A_k^c})_{k\in\mathbb{N}} \text{ is 1-equi-integrable} \end{array}\right.$$

Proof of Lemma IV.2.9. For all $\epsilon > 0$ and $n \in \mathbb{N}$, we define

$$\left[\begin{array}{l} \delta_n(\epsilon) = \text{Sup} \left\{ \int_A |f_n(t)|\, dt / \lambda(A) \le \epsilon \right\} \\ \delta(\epsilon) = \lim\sup_n \delta_n(\epsilon) \end{array}\right.$$

δ is a decreasing function of ϵ which tends to some $\delta \ge 0$ when ϵ tends to 0.

If $\epsilon_k \xrightarrow[k\to+\infty]{} 0$, then $\delta(\epsilon_k) \xrightarrow[k\to+\infty]{} \delta$ and we can choose a sequence $(n_k)_{k\in\mathbb{N}}$ such that $\delta_{n_k}(\epsilon_k) \xrightarrow[k\to+\infty]{} \delta$.

For all $k \in \mathbb{N}$ there exists $A_k \in \mathscr{B}$ such that

$$\left[\begin{array}{l} \lambda(A_k) \le \epsilon_k \\ \int_{A_k} |f_{n_k}(t)|\, dt \ge \delta_{n_k}(\epsilon_k) - \epsilon_k \end{array}\right.$$

Hence:

$$\int_{A_k} |f_{n_k}(t)|\, dt \xrightarrow[k\to+\infty]{} \delta$$

Let us prove that $(f_{n_k}1_{A_k^c})_{k\in\mathbb{N}}$ is 1-equi-integrable: If not, there exist $\alpha > 0$, a subsequence $(f_{n_{k_p}})_{p\in\mathbb{N}}$, and $(B_p)_{p\in\mathbb{N}} \subset \mathscr{B}$ such that

$$\left[\begin{array}{l} \lambda(B_p) \xrightarrow[p\to+\infty]{} 0 \\ \int_{B_p} |1_{A_{k_p}^c}(t)f_{n_{k_p}}(t)|\, dt \ge \alpha > 0 \end{array}\right.$$

Then $(\int_{A_{k_p}\cup B_p} |f_{n_{k_p}}(t)|\, dt)_{p\in\mathbb{N}}$ converges when $p \to +\infty$ to something which is more than $\delta + \alpha$ and $(P(A_{k_p} \cup B_p))_{p\in\mathbb{N}}$ converges to 0. This contradicts the definition of δ.

Proof of Theorem IV.2.8. We just have to apply Lemma IV.2.9 with $|f_n|^p$ and take

$$g_k = |f_{n_k}|^p 1_{A_k^c}, \qquad h_k = |f_{n_k}|^p 1_{A_k}$$

Definition IV.2.10. Any sequence $(h_k)_{k \in \mathbb{N}}$ in $L_p[0,1]$ which is bounded and such that $\lambda(\text{Supp } h_k) \to_{k \to 0} 0$ is called a *peak sequence*.

Remark. It is an easy exercise to show that if a peak sequence in L_p does not converge in norm to 0, it has a subsequence which is equivalent to the unit vector basis of ℓ_p.

In order to study subspaces of L_p-spaces, we need to know their *Rademacher type and cotype*. Let us give basic tools and definitions related to these notions (see [P2], [P3], [P4], [LT]).

First of all, we want to recall the definition of Rademacher functions on [0,1], which was already given in Definition I.3.7.

Definition IV.2.11. The sequence of *Rademacher functions* $(r_n)_{n \in \mathbb{N}}$ on [0,1] is defined by

$$r_0 = 1$$

$$r_1 = 1 \quad \text{on } [0,1/2[$$

$$\quad = -1 \text{ on } [1/2,1]$$

$$\vdots$$

$$\text{For } n \in \mathbb{N}, \quad r_n = 1 \quad \text{on } \bigcup_{k=0}^{2^{n-1}-1} \left[\frac{2k}{2^n}, \frac{2k+1}{2^n} \right[$$

$$\quad = -1 \text{ on } \bigcup_{k=0}^{2^{n-1}-1} \left[\frac{2k+1}{2^n}, \frac{2k+2}{2^n} \right]$$

$$\vdots$$

Remark. The sequence $(r_n)_{n \in \mathbb{N}}$ is a sequence of independent and identically distributed (i.i.d.) random variables, which is bounded in $L_p[0,1]$ for all $p \in [1, +\infty]$.

We recall two well-known inequalities, which can be found in [LT] or [Be3]. Here, we are going to give a probalistic proof of them, which is very short and which is due to J. Hoffmann-Jorgensen [Ho]:

Proposition IV.2.12

Khintchine's inequalities: For every $1 \leq p < +\infty$, there exist $A_p, B_p > 0$ such that for all $n \in \mathbb{N}$ and $(a_i)_{i=0,\dots,n} \in \mathbb{R}^{n+1}$:

$$A_p \left(\sum_{i=0}^{n} |a_i|^2 \right)^{1/2} \leq \left(\int_0^1 \left| \sum_{i=0}^{n} r_i(t)a_i \right|^2 dt \right)^{1/2} \leq B_p \left(\sum_{i=0}^{n} |a_i|^2 \right)^{1/2}$$

We will note also in short:

$$\left(\int_0^1 \left| \sum_{i=0}^{n} r_i(t)a_i \right|^2 dt \right)^{1/2} \sim \left(\sum_{i=0}^{n} |a_i|^2 \right)^{1/2}$$

Khintchine-Kahane's inequalities: For every $0 < p < q < \infty$, there exists $K_{pq} > 0$ such that for all Banach spaces X, $n \in \mathbb{N}$ and $(x_i)_{i=0,\dots,n} \in X^{n+1}$:

$$\left(\int_0^1 \left\| \sum_{i=0}^{n} r_i(t)x_i \right\|_X^q dt \right)^{1/q} \leq K_{p,q} \left(\int_0^1 \left\| \sum_{i=0}^{n} r_i(t)x_i \right\|_X^p dt \right)^{1/p}$$

We will note also in short:

$$\left(\int_0^1 \left\| \sum_{i=0}^{n} r_i(t)x_i \right\|_X^q dt \right)^{1/q} \sim \left(\int_0^1 \left\| \sum_{i=0}^{n} r_i(t)x_i \right\|_X^p dt \right)^{1/p}$$

Proof. Since $\overline{\text{Span}} \, [r_i, i \in \mathbb{N}]$ is obviously isometric to ℓ_2 in L_2, it is sufficient to show Khintchine-Kahane's inequalities:

Let $x_0, x_1, \dots, x_n$ be given in the Banach space X and define for all $k \leq n$, $S_k = \sum_{i=0}^{k} r_i x_i$, $R_k = \sum_{i=k+1}^{n} r_i x_i$.

S_k and R_k are random variables on $[0,1]$ with values in X.

We want to show that the norms of $\| S_n \|_X$ in L_p and in L_q are equivalent. We need a maximal inequality which is due to P. Levy:

Lemma IV.2.13. For all $t > 0$, we have

$$\lambda \{ \underset{0 \leq k \leq n}{\text{Max}} \, \| S_k \|_X > t \} \leq 2\lambda \{ \| S_n \|_X > t \}$$

Proof of Lemma IV.2.13. For all $k \leq n$, define the set A_k by

$$A_k = \{ \| S_k \|_X > t, \, \underset{0 \leq j < k}{\text{Max}} \, \| S_j \|_X \leq t \}$$

The set $\{ \text{Max}_{0 \leq k \leq n} \, \| S_k \|_X > t \}$ is the disjoint union of the A_k's. So we can write, since $\{\| S_n \|_X > t\} \subseteq \{\text{Max}_{0 \leq k \leq n} \| S_k \|_X > t\}$,

(1) $$\lambda\{\| S_n \|_X > t\} = \sum_{k=0}^{n} \lambda(\{\| S_n \|_X > t\} \cap A_k)$$

Since the Rademacher functions are symmetric, that is, the law of every random variable r_k is the same as the law of $-r_k$, and since the sequence of Rademacher functions is independent, the law of $(r_0, \ldots, r_k, -r_{k+1}, \ldots -r_n)$ is the same as the law of $(r_0, \ldots, r_k, r_{k+1}, \ldots r_n)$.

That implies that the random variables $S_k - R_k$ have the same law as the random variable $S_n = S_k + R_k$ for all $k \leq n$. Thus we can write:

(2) $$\lambda\{\| S_n \|_X > t\} = \sum_{k=0}^{n} \lambda(\{\| S_k - R_k \|_X > t\} \cap A_k)$$

Writing $2S_k = S_n + S_k - R_k$, we get on A_k

$$2t < 2 \| S_k \|_X \leq \| S_n \|_X + \| S_k - R_k \|_X$$

This means that, on A_k, either $\| S_n \|_X > t$ or $\| S_k - R_k \|_X > t$. Thus:

$$\lambda(A_k) \leq \lambda(\{\| S_n \|_X > t\} \cap A_k) + \lambda(\{\| S_k - R_k \|_X > t\} \cap A_k)$$

Summing these inequalities for $k = 0, 1, \ldots, n$, we get, by (1) and (2),

$$\lambda\{\underset{0 \leq k \leq n}{\text{Max}} \| S_k \|_X > t\} = \sum_{k=0}^{n} \lambda(A_k) \leq 2\lambda\{\| S_n \|_X > t\}$$

This proves Lemma IV.2.13.

We come back to the proof of Proposition IV.2.12: Let us choose $t_0 > 0$ such that

(3) $$\lambda\{\| S_n \|_X > t_0\} < 1/2$$

Then $\text{Max}_{0 \leq i \leq n} \| x_i \| \leq 2t_0$, because otherwise, writing $\| x_i \| = \| S_i - S_{i-1} \|_X$, we would get $\text{Max}_{0 \leq i \leq n} \| S_i \|_X > t_0$ and thus, by Lemma IV.2.13,

$$\lambda\{\| S_n \|_X > t_0\} \geq 1/2$$

which contradicts (3).

For all $t \geq t_0$, we can then write, with the notation of Lemma IV.2.13,

$$\lambda\{\|\, S_n \,\|_X > 5t\} = \sum_{k=0}^{n} \lambda(\{\|\, S_n \,\|_X > 5t\} \cap A_k)$$

By definition, on A_k we have

$$\|\, S_n \,\|_X \leq \|\, S_{k-1} \,\|_X + \|\, x_k \,\| + \|\, S_n - S_k \,\|_X$$
$$\leq t + 2t_0 + \|\, S_n - S_k \,\|_X$$
$$\leq 3t + \|\, S_n - S_k \,\|_X$$

Thus, we get

$$\lambda\{\|\, S_n \,\|_X > 5t\} = \sum_{k=0}^{n} \lambda(\{\|\, S_n - S_k \,\|_X > 2t\} \cap A_k)$$

Since $S_n - S_k$ and A_k are independent, we can write, for all $k = 0, 1,$
$\ldots$ n,

$$\lambda(\{\|\, S_n - S_k \,\|_X > 2t\} \cap A_k) = \lambda\{\|\, S_n - S_k \,\|_X > 2t\} \lambda(A_k)$$
$$\leq \lambda\{\|\, S_n \,\|_X + \|\, S_k \,\|_X > 2t\} \lambda(A_k)$$
$$\leq (\lambda\{\|\, S_n \,\|_X > t\} + \lambda\{\|\, S_k \,\|_X > t\}) \lambda(A_k)$$
$$\leq 3\lambda\{\|\, S_n \,\|_X > t\} \lambda(A_k)$$

by Lemma IV.2.13. Thus, for $t \geq t_0$,

$$\lambda\{\|\, S_n \,\|_X > 5t\} \leq 3\lambda\{\|\, S_n \,\|_X > t\} \sum_{k=0}^{n} \lambda(A_k)$$

(4)
$$\leq 3\lambda\{\|\, S_n \,\|_X > t\} \lambda\{\underset{0 \leq k \leq n}{\text{Max}} \|\, S_k \,\|_X > t\}$$

$$\leq 6\lambda\{\|\, S_n \,\|_X > t\}^2$$

again using Lemma IV.2.13.

Let p and q be two given positive numbers. Choose t_0 such that

$$t_0^q = 12 \; 5^p \int \|\, S_n \,\|_X^q \, d\lambda$$

Obviously, we get by Tchebychev's inequality (see Proposition IV.2.3):

$$\lambda\{\|\, S_n \,\|_X > t_0\} \leq 1/t_0^q \int \|\, S_n \,\|_X^q \, d\lambda \leq 1/12 \; 5^p$$

In particular, condition (3) is satisfied.

Let us compute $\int \| S_n \|_X^p \, d\lambda$ by the formula of integration by parts:

$$\int \| S_n \|_X^p \, d\lambda = 5^p \int_0^\infty \lambda\{\| S_n \|_X > 5t\} \, p t^{p-1} \, dt$$

$$\leq (5t_0)^p + 5^p \int_{t_0}^\infty \lambda\{\| S_n \|_X > 5t\} \, p t^{p-1} \, dt$$

and by (4):

$$\int \| S_n \|_X^p \, d\lambda \leq (5t_0)^p + 6 \cdot 5^p \int_{t_0}^\infty \lambda\{\| S_n \|_X > t\}^2 \, p t^{p-1} \, dt$$

$$\leq (5t_0)^p + (1/2) \int_{t_0}^\infty \lambda\{\| S_n \|_X > t\} \, p t^{p-1} \, dt$$

because, by the choice of t_0, for all $t \geq t_0$,

$$\lambda\{\| S_n \|_X > t\} \leq \lambda\{\| S_n \|_X > t_0\} \leq 1/12 \cdot 5^p$$

We thus get

$$\int \| S_n \|_X^p \, d\lambda \leq (5t_0)^p + (1/2) \int \| S_n \|_X^p \, d\lambda$$

This implies

$$\int \| S_n \|_X^p \, d\lambda \leq 2(5t_0)^p$$

It is enough now to apply the definition of t_0 to obtain

$$\left(\int \| S_n \|_X^p \, d\lambda \right)^{1/p} \leq 2^{1/p} \, 5 \, (12 \cdot 5^p)^{1/q} \left(\int \| S_n \|_X^q \, d\lambda \right)^{1/q}$$

This finishes the proof of Proposition VI.2.12.

Definition IV.2.14. A Banach space X is *of Rademacher type p* $(1 \leq p \leq 2)$ if there exists $T_p > 0$ such that for all $n \in \mathbb{N}$ and $(x_i)_{i=0,\ldots,n} \in \mathbb{R}^{n+1}$:

$$\left(\int_0^1 \left\| \sum_{i=0}^n r_i(t)x_i \right\|_X^2 \, dt \right)^{1/2} \leq T_p \left(\sum_{i=0}^n \| x_i \|_X^p \right)^{1/p}$$

The Banach space X is of *Rademacher cotype q* $(2 \leq q \leq +\infty)$ if there exists $C_q > 0$ such that for all $n \in \mathbb{N}$ and $(x_i)_{i=0,\ldots,n} \in \mathbb{R}^{n+1}$:

$$C_q \left(\sum_{i=0}^n \| x_i \|_X^q \right)^{1/q} \leq \left(\int_0^1 \left\| \sum_{i=0}^r r_i(t)x_i \right\|_X^2 \, dt \right)^{1/2}$$

(If $q = \infty$, replace $(\sum_{i=0}^n \| x_i \|_X^q)^{1/q}$ by $\text{Sup}_{0 \leq i \leq n} \| x_i \|_X$.)

T_p is called the *Rademacher type p constant* and C_q the *Rademacher cotype q constant*.

It is clear that every Banach space is of Rademacher type 1 and Rademacher cotype ∞. If X is of Rademacher type p (resp. Rademacher cotype q) it is of Rademacher type p' for all p' in $[1,p]$ (resp. of Rademacher cotype q' for all q' in $[q,+\infty]$).

These properties are stable by isomorphism and pass to subspaces.

Theorem IV.2.15. For $1 \le p \le 2$, ℓ_p and $L_p[0,1]$ are of Rademacher type p and Rademacher cotype 2. For $2 \le p < \infty$, ℓ_p and $L_p[0,1]$ are of Rademacher type 2 and Rademacher cotype p. Moreover, these values are the best possible.

Proof. In order to give only one proof for ℓ_p and $L_p[0,1]$, here we are going to work on an infinite-dimentional $L_p(\Omega,\Sigma,\mu)$ for any measure μ.

For $(x_i)_{i=0,\ldots,n}$ in $L_p = L_p(\Omega,\Sigma,\mu)$ we can write, using first Khintchine-Kahane's inequalities and then Khintchine's inequalities,

$$\left(\int_0^1 \left\| \sum_{i=0}^n r_i(t)x_i \right\|_{L_p}^2 dt\right)^{1/2} \sim \left(\int_0^1 \left\| \sum_{i=0}^n r_i(t)x_i \right\|_{L_p}^p dt\right)^{1/p}$$

$$= \left(\int_0^1 \int_\Omega \left| \sum_{i=0}^n r_i(t)x_i(\omega) \right|^p d\mu(\omega)\, dt\right)^{1/p}$$

$$= \left(\int_\Omega \int_0^1 \left| \sum_{i=0}^n r_i(t)x_i(\omega) \right|^p dt\, d\mu(\omega)\right)^{1/p}$$

$$\sim \left(\int_\Omega \left(\sum_{i=0}^n |x_i(\omega)|^2 \right)^{p/2} d\mu(\omega)\right)^{1/p}$$

Suppose that $1 < p < 2$. (The case where $p \ge 2$ is analogous.) In $L_{p/2}(\Omega,\Sigma,\mu)$, by Lemma I.3.10 we have $\| x + y \|_{p/2} \ge \| x \|_{p/2} + \| y \|_{p/2}$. Then

$$\left(\int_\Omega \left(\sum_{i=0}^n |x_i(\omega)|^2 \right)^{p/2} d\mu(\omega)\right)^{1/p} \ge \left(\sum_{i=0}^n \left(\int_\Omega |x_i(\omega)|^p d\mu(\omega)\right)^{2/p} \right)^{1/2}$$

$$= \left(\sum_{i=0}^n \| x_i \|_p^2 \right)^{1/2}$$

and this proves that $L_p(\Omega,\Sigma,\mu)$ is of Rademacher cotype 2, which obviously is the best possible.

On the other hand, we use the fact that the norm of ℓ_2 is less than the norm of ℓ_p. Then

$$\left(\int_\Omega \left(\sum_{i=0}^n |x_i(\omega)|^2\right)^{p/2} d\mu(\omega)\right)^{1/p} \le \left(\int_\Omega \sum_{i=0}^n |x_i(\omega)|^p\right)^{1/p}$$

$$= \left(\sum_{i=0}^n \|x_i\|_p^p\right)^{1/p}$$

This proves that $L_p(\Omega,\Sigma,\mu)$ is of Rademacher type p.

To prove that p is the best possible Rademacher type, it is sufficient to show that ℓ_p is not of Rademacher type $p + \epsilon$ for $\epsilon > 0$ (because $L_p(\Omega,\Sigma,\mu)$ has ℓ_p as a subspace; see Corollary IV.1.6). But, if $(e_n)_{n\in\mathbb{N}}$ is the unit vector basis of ℓ_p, we get

$$\begin{cases} \left(\int_0^1 \left\|\sum_{i=0}^n r_i(t)e_i\right\|_{\ell_p}^p dt\right)^{1/p} = (n+1)^{1/p} \\ \left(\sum_{i=0}^n \|x_i\|^{p+\epsilon}\right)^{1/p+\epsilon} = (n+1)^{1/p+\epsilon} \end{cases}$$

Letting $n \to +\infty$ proves our assumption.

3. ULTRAPOWERS OF L_p-SPACES

Results of this section are based on the representation of abstract L_p-spaces due to S. Kakutani [Ka]. This important result is given with various extensions in [LT] volume II, [H], [K1] and [L].

Theorem IV.3.1. Let $1 \le p < +\infty$. Every ultrapower of ℓ_p or $L_p[0,1]$ is an L_p-space.

Before proving this theorem, we have to recall some properties of *lattices*:

Definition IV.3.2. A real Banach space X is called a *lattice* if there is a partial order $\le$ on X which verifies:

(i) $x \le y \Rightarrow x + y \le y + z$.

   (ii) For all real numbers $a > 0$ and all positive elements x in X, the element ax is positive in X.

  (iii) All elements x and y in X have a *supremum* and an *infimum*, respectively $x \vee y$ and $x \wedge y$. Moreover, $x + y = x \vee y + x \wedge y$.

  (iv) If $|x| \leq |y|$, then $\|x\| \leq \|y\|$, where $|x|$ denotes the element $x \vee (-x)$.

It is clear that any L_p-space is a lattice for the natural order of function classes.

We will also consider an additional property of lattices called $(*_p)$:

$$(*_p) \quad \begin{cases} \text{if } x \geq 0 \text{ and } y \geq 0, \ \|x + y\|_p^p \geq \|x\|_p^p + \|y\|_p^p \geq \|x \vee y\|_p^p \\ \| \, |x| \, \|_p = \|x\|_p \end{cases}$$

It is obvious that the L_p-norm has this property $(*_p)$

First of all, we are going to prove that these lattice properties pass to ultrapowers:

Lemma IV.3.3. Let X be a Banach lattice with $(*_p)$. Then any ultrapower $\tilde{X} = X^I/\mathcal{U}$ of X is a Banach lattice with $(*_p)$.

Proof of Lemma IV.3.3. On $\tilde{X} = X^I/\mathcal{U}$, we define a partial order by

$$\tilde{x} \leq \tilde{y} \Leftrightarrow \exists (x_i)_{i \in I}, \ (y_i)_{i \in I} \in X^I \times X^I \text{ such that } \begin{cases} \tilde{x} = (x_i)_{i \in I} \\ \tilde{y} = (y_i)_{i \in I} \\ \forall i \in I, \ x_i \leq y_i \end{cases}$$

Then it is easy to verify that $\tilde{X}$ with this order is a Banach lattice with property $(*_p)$.

To prove Theorem IV.3.1 it is thus enough to prove the following *representation of abstract L_p-spaces*, which is due to S. Kakutani [Ka]:

Theorem IV.3.4. If X is a Banach lattice with $(*_p)$, then there exists a measure space (Ω, Σ, μ) such that $X = L_p(\Omega, \Sigma, \mu)$.

Proof of Theorem IV.3.4. Remark first that if X has $(*_p)$, then for all $x, y \in X$, $x \geq 0$, $y \geq 0$ such that $x \wedge y = 0$, we get $x + y = x \vee y$ and thus $\|x + y\|^p = \|x\|^p + \|y\|^p$.

The proof of this theorem is long and will be divided into six steps:

First step: Let us show that every increasing and order-bounded sequence in X converges:

Indeed, suppose that $(x_n)_{n \in \mathbb{N}}$ is an increasing sequence in X and that there exists x in X such that $\forall n \in \mathbb{N}$, $x_n \leq x$. Then $(x - x_n)_{n \in \mathbb{N}}$ is decreasing and positive. By $(*_p)$ we get, for $(n,m) \in \mathbb{N}^2$,

$$\| x - x_n \|^p \geq \| x - x_m \|^p + \| x_n - x_m \|^p \geq 0$$

Since $(\| x - x_n \|)_{n \in \mathbb{N}}$ is decreasing, it converges to some $\alpha \geq 0$. Thus $(\| x_n - x_m \|^p)_{n,m \in \mathbb{N}}$ converges to 0 when n and m tend to $+\infty$. This proves that $(x_n)_{n \in \mathbb{N}}$ is a Cauchy sequence in X. Thus it converges.

Second step: For all $e \in X$, $e \geq 0$, we define

$$Z(e) = \{x \in X / \exists a \in \mathbb{R}^+, ae \geq | x |\}.$$

It is clear that $Z(e)$ is a sublattice of X. If $(e_0, \ldots, e_k)$ are strictly positive in X and such that $e_i \wedge e_j = 0$ for $i \neq j$, let us prove that $Z(e_0), \ldots, Z(e_k)$ are linearly independent subspaces of X:

If these spaces were not independent, we could find $x_0 \in Z(e_0)$, $\ldots, x_k \in Z(e_k)$, where at least one x_i is nonzero such that $\sum_{i=0}^{k} x_i = 0$. Suppose for instance that $x_0 \neq 0$. We get

$$| x_0 | \leq (| x_1 | + \cdots + | x_k |)$$

By definition, for all $i = 0, \ldots, k$, there exist $a_i \in \mathbb{R}^+$ such that

$$a_i e_i \geq | x_i |$$

If $A = \text{Sup}\{a_i, i = 1, \ldots, k\}$, we get

$$| x_0 | \leq A(e_1 + \cdots + e_k)$$

and thus

$$| x_0 | \leq \text{Sup}\{a_0, A\}[e_0 \wedge (e_1 + \cdots + e_k)]$$

Since $(e_0, e_1, \ldots, e_k)$ are disjoint, this implies $x_0 = 0$ and contradicts our hypothesis.

Third step: Let us consider the family of subsets E of X such that

$$\begin{bmatrix} \text{(i) } \forall x \in E, x \geq 0 \text{ and } \| x \| = 1 \\ \text{(ii) } \forall (x,y) \in E^2, x \neq y \Rightarrow x \wedge y = 0 \end{bmatrix}$$

Let E_0 be a maximal element of this family, ordered by inclusion. (The existence of E_0 is obvious by Zorn's lemma.) Then let us prove that the subspace $Z = \text{Span}\{Z(e), e \in E_0\}$ is dense in X.

Since every element of X is the difference of two positive elements, it is sufficient to show that if x is positive in X, then $x \in \overline{Z}$. Let $F_0 \subset E_0, |F_0| < +\infty$. Then

$$x \geq \bigvee_{e \in F_0} (x \wedge e) = \sum_{e \in F_0} (x \wedge e)$$

By $(*_p)$, we can write

$$\| x \|^p \geq \| \sum_{e \in F_0} x \wedge e \|^p \geq \sum_{e \in F_0} \| x \wedge e \|^p$$

This inequality is true for all $F_0 \subset E_0, |F_0| < +\infty$.

This proves that the family $(\| x \wedge e \|^p)_{e \in E_0}$ is convergent. We deduce from this that the set $\{e \in E_0 / x \wedge e \neq 0\}$ is countable. Denote this set by $\{e_0, e_1, \ldots, e_n, \ldots\}$ and define

$$f = \sum_{k=0}^{\infty} \frac{e_k}{2^k}$$

Then $f \in X$ and for $n \in \mathbb{N}$, we have

$$x \wedge nf = \sum_{k=0}^{\infty} x \wedge \frac{ne_k}{2^k}$$

Since $x \wedge ne_k/2^k \in Z(e_k)$, this implies that $x \wedge nf \in \overline{Z}$. The sequence $(x \wedge nf)_{n \in \mathbb{N}}$ is increasing and majorized by x. By the first step, it has a limit $u \in \overline{Z}$. Let us prove that $x = u$:

Set $v = x - u$ and $w = v \wedge f$. For all $n \in \mathbb{N}$, we have

$$0 \leq w \leq v \leq x - (x \wedge nf)$$

Thus

$$w + (x \wedge nf) \leq x$$

By induction, we are going to prove that

$$\forall k \in \mathbb{N}, \quad kw \leq x$$

It is true for $k = 0$. Suppose that it is true for one $k \in \mathbb{N}$, that is, $kw \leq x$. Then

$$w + (x \wedge kf) \leq x$$
$$\Rightarrow w + (kw \wedge kf) \leq x$$
$$\Rightarrow (k + 1)w \leq x \quad \text{since } w \leq f$$

So it is true for $k + 1$ and this proves our assumption.

We deduce from this that $k \| w \| \leq \| x \|$ for all $k \in \mathbb{N}$. Thus $w = 0$. That means that $v \wedge f = 0$ and thus

$$\forall n \in \mathbb{N}, \qquad v \wedge e_n = 0$$

This implies also by density that $v \wedge e = 0$ for all $e \in E_0$. If $v \neq 0$, then $E_0 \cup v/\| v \|$ would have properties (i) and (ii), which is impossible by the maximality of E_0. Thus $v = 0$, and $x = u \in \overline{Z}$.

Fourth step: Suppose that we have already proved the existence, for all e in E_0, of a measure space $(\Omega_e, \Sigma_e, \mu_e)$ such that $\overline{Z}(e)$ is isometric to $L_p(\Omega_e, \Sigma_e, \mu_e)$. Then we are going to show that there exists a measure space (Ω, Σ, μ) such that $X = \overline{Z}$ is isometric to $L_p(\Omega, \Sigma, \mu)$.

Define

$$\Omega = \prod_{e \in E_0} \Omega_e, \qquad \Sigma = \bigotimes_{e \in E_0} \Sigma_e, \qquad \mu = \bigoplus_{e \in E_0} \mu_e$$

That means that a k-tuple $A = (A_{e_0}, \ldots, A_{e_k})$ belongs to Σ if $A_{e_i} \in \Sigma_{e_i}$ for all $i = 0, \ldots, k$ and then $\mu(A) = \sum_{i=0}^{k} \mu_{e_i}(A_{e_i})$.

For all $e \in E_0$, let φ_e be an isometry from $\overline{Z}(e)$ onto $L_p(\Omega_e, \Sigma_e, \mu_e)$. Let x belong to Z. Then there exist $x_0 \in Z(e_0), \ldots, x_k \in Z(e_k)$ such that $x = x_0 + \cdots + x_k$. Then by definition $| x_i | \wedge | x_j | = 0$ for $i \neq j$. Thus $\| x_0 + \cdots + x_k \|^p = \| x_0 \|^p + \cdots + \| x_k \|^p$ by $(*_p)$. If we set $\varphi(x) = \sum_{i=0}^{k} \varphi_{e_i}(x_i)$, this defines an isometry on Z which can be extended to $\overline{Z} = X$ and by construction $\varphi(X) = L_p(\Omega, \Sigma, \mu)$.

Fifth step: For fixed $e \in E_0$, we are going to define a measure space $(\Omega_e, \Sigma_e, \mu_e)$ such that $\overline{Z}(e)$ is isometric to $L_p(\Omega_e, \Sigma_e, \mu_e)$.

Define $S = \{u \in Z(e)/u \wedge (e - u) = 0\}$ and let us give some properties of S:

Proposition IV.3.5

1. $u \in S \Rightarrow 0 \leq u \leq e$ and $e - u \in S$.
2. $u, v \in S \Rightarrow u \vee v \in S$.
3. If $(u_n)_{n \in \mathbb{N}} \subset S$ is increasing then $(u_n)_{n \in \mathbb{N}}$ converges in S.

Proof of Proposition IV.3.5

1. is obvious.
2. We can write, for $u, v \in S$,

$$e - (u \vee v) = -[(u - e) \vee (v - e)] = (e - u) \wedge (e - v)$$

Thus

$$(u \vee v) \wedge [e - (u \vee v)] = (u \vee v) \wedge [(e - u) \wedge (e - v)]$$
$$= [u \wedge (e - u) \wedge (e - v)] \vee [v \wedge (e - v) \wedge (e - u)] = 0$$

This proves that $u \vee v \in S$.

3. If $(u_n)_{n \in \mathbb{N}}$ is an increasing sequence in S, then, since $0 \leq u_n \leq e$ for all $n \in \mathbb{N}$, it converges to some u in X by the first step. To prove that u belongs to S, it is enough to write

$$u \wedge (e - u) = \lim_n u_n \wedge (e - u_n) = 0$$

Proposition IV.3.5 shows that $(S, \wedge, \vee)$ is a σ-algebra.

This step is now achieved by using the following well-known *representation theorem* due to M. H. Stone [Sto]:

Every σ-algebra with a unit is isomorphic to the Boolean algebra of all simultaneously open and closed subsets of a totally disconnected compact Hausdorff space Ω.

A proof of this classical result can be found in [DS], I.2.1, or in [L].

We can define a measure μ on S by

$$\forall u \in S, \qquad \mu(u) = \| u \|^p$$

μ *is additive:* Indeed if $u \wedge v = 0$,

$$\| u \vee v \|^p = \| u \|^p + \| v \|^p \text{ by } (*_p)$$

μ *is σ-additive:* Indeed if $(u_n)_{n \in \mathbb{N}}$ is increasing in S, by Proposition IV.3.5.3, $(u_n)_{n \in \mathbb{N}}$ converges to some u in S. Thus $(\| u_n \|^p)_{n \in \mathbb{N}}$ converges to $\| u \|^p$.

If $\mathscr{E}$ is the linear space generated by S in $Z(e)$, then $\mathscr{E}$ is isometric by definition to the space of simple functions on (Ω, S, μ) equipped with the L_p-norm. To prove that $L_p(\Omega, S, \mu)$ is isometric to $\overline{Z(e)}$, it remains to show that $\mathscr{E}$ is dense in $Z(e)$.

Sixth step: Let us show that $\mathscr{E}$ is dense in $Z(e)$. We need a lemma.

Lemma IV.3.6. Let $x \in Z(e)$, $x \geq 0$ and $x \neq 0$. Then there exist $u \in S$, $u \neq 0$, and $n \in \mathbb{N}$ such that $x \geq u/n$.

Proof of Lemma IV.3.6. By multiplying x by some integer, we can suppose that x is not less than e. Define

$$\begin{cases} z = (x - e) \vee 0 \\ u = \lim_n e \wedge nz \end{cases}$$

Since z is positive, u is also positive. If u were null, this would imply that $e \wedge z = 0$. But $z \in Z(e)$, so there exists $p \in \mathbb{R}^+$ such that $pe \geq z$. Thus

$$0 = p(e \wedge z) \geq pe \wedge z \geq z \wedge z = z$$

Since $z \neq 0$ this implies that $u \neq 0$.

First of all, let us prove that $u \in S$: For $k \in \mathbb{N}$, set $w_k = (e - u) \wedge kz$. Then

$$0 \leq w_k \leq kz$$
$$0 \leq w_k \leq (e - u) \leq e - (e \wedge nz) \quad \text{for all } n \in \mathbb{N}.$$

Let us prove that $w_k = 0$ for all $k \in \mathbb{N}$. Indeed, if we suppose that $nw_k \leq e$ for some $n \in \mathbb{N}$, then we can write

$$w_k \leq e - (e \wedge nkz)$$
$$\Rightarrow w_k + (e \wedge nkz) \leq e$$
$$\Rightarrow w_k + (nw_k \wedge nkz) \leq e$$
$$\Rightarrow w_k + n(w_k \wedge kz) \leq e$$
$$\Rightarrow (n + 1)w_k \leq e$$

Thus, since this property is true for $n = 0$ we get, for all n in $\mathbb{N}$, $nw_k \leq e$. This implies $w_k = 0$. Thus $(e - u) \wedge kz = 0 = (e - u) \wedge (e \wedge kz)$. When $k \to +\infty$, this implies $(e - u) \wedge u = 0$ and $u \in S$.

Let us show that $x \geq u$: Set $w = e \wedge n(x - e)$. We have

$$w \leq e$$
$$w \leq nx - ne$$

Thus we can write

$$w + ne \leq nx$$
$$\Rightarrow (n + 1)w \leq nx$$
$$\Rightarrow \left(1 + \frac{1}{n}\right) w \leq x$$

And since by hypothesis x is positive, we get

$$x \geq \left(1 + \frac{1}{n}\right) w \vee 0 = \left(1 + \frac{1}{n}\right)(w \vee 0) \geq (w \vee 0)$$

Replacing w by its value, we get

$$x \geq [e \wedge n(x - e)] \vee 0 = e \wedge [n(x - e) \vee 0] = e \wedge nz$$

When $n \rightarrow +\infty$, this last inequality gives $x \geq u$. This finishes the proof of Lemma IV.3.6.

We come back to the proof of the sixth step. Suppose that $\mathscr{E}$ is not dense in $Z(e)$. Then there exist $x \in Z(e)$, $x \geq 0$ such that $x \notin \overline{\mathscr{E}}$. Set $a = \text{Sup } \{\| y \|^p / y \in \overline{\mathscr{E}} \text{ and } 0 \leq y \leq x\}$. There exists an increasing sequence $(y_n)_{n \in \mathbb{N}}$ in $\overline{\mathscr{E}}$ such that

$$0 \leq y_n \leq x$$
$$\| y_n \|^p \xrightarrow[n \rightarrow +\infty]{} a$$

By the first step, $(y_n)_{n \in \mathbb{N}}$ converges to some $y \in \overline{\mathscr{E}}$ such that

$$\| y \|^p = a$$
$$0 \leq y \leq x$$

Since $x \notin \overline{\mathscr{E}}$, $x - y \neq 0$, and by Lemma IV.3.6 there exist $u \in S$, $u \geq 0$, $u \neq 0$, and $n \in \mathbb{N}$ such that $x - y \geq u/n$. But we have

$$y + \frac{u}{n} \in \overline{\mathscr{E}}$$

$$0 \leq y + \frac{u}{n} \leq x$$

$$\| y + \frac{u}{n} \|^p \geq \| y \|^p + \| \frac{u}{n} \|^p > a \quad \text{by } (*_p)$$

Thus, we get a contradiction and this proves that $\overline{\mathscr{E}} = Z(e)$.

This last step finishes the proof of Theorem IV.3.4 and thus of Theorem IV.3.1.

This representation of ultrapowers of L_p-spaces has the following obvious corollary:

Corollary IV.3.7. For all $p \in [1, +\infty[$, any L_p-space is superstable and for all $p \in]1, +\infty[$, any L_p-space is superreflexive.

The next result is a consequence of Theorems IV.3.1 and II.5.13.

Corollary IV.3.8. For $1 \leq p,q < +\infty$, the following properties are equivalent:
    (i) ℓ_q is isomorphic to a subspace of L_p.
    (ii) ℓ_q is isometric to a subspace of L_p.
    (iii) ℓ_q is f.r. in L_p.

4. ℓ_q SUBSPACES OF $L_p[0,1]$

First, it is important to note that c_0 is not isomorphic to any subspace of $L_p[0,1]$ for $1 \leq p < +\infty$. When $p > 1$, this is because c_0 is not reflexive. When $p = 1$, we already proved the result in Lemma I.3.9.

The first two theorems are easy and well known:

Theorem IV.4.1. For all $p \in [1, +\infty[$, ℓ_2 is isometric to a subspace of $L_p[0,1]$.

Proof. Let us recall that a *standard gaussian random variable* (in short, *gaussian r.v.*) is defined by its density on $\mathbb{R}$, namely $1/(2\pi)^{1/2} e^{-t^2/2}$. It is well known and easy to prove that such a r.v. belongs to $L_p[0,1]$ for all $p \in [1, +\infty]$.

As usual, we will denote by *i.i.d.* an *independent and identically distributed* sequence of r.v.

An introduction to these classical notions can be found in [Fe].

Let $(g_n)_{n \in \mathbb{N}}$ be a sequence of i.i.d gaussian random variables on $[0,1]$. Then by definition, the density of $\sum_{i=0}^{\infty} a_i g_i$ is $1/(2\pi)^{1/2}$ exp $(\sum_{i=0}^{\infty} a_i^2 t^2/2)$. Thus we can write, for all $(a_i)_{n \in \mathbb{N}} \subset \mathbb{R}^{(\mathbb{N})}$,

$$\left\| \sum_{i=0}^{\infty} a_i g_i \right\|_p = \left(\sum_{i=0}^{\infty} a_i^2 \right)^{1/2} \| g_1 \|_p$$

Since $\| g_1 \|_p < +\infty$ for all $p \in [1, +\infty[$, this proves that $\overline{\text{Span}} [g_n, n \in \mathbb{N}]$ is isometric to ℓ_2 in $L_p[0,1]$.

Another way to prove Theorem IV.4.1. is to use Khintchine's inequality (see Proposition IV.2.12) which proves that $\overline{Span} \ [r_n, \ n \in \mathbb{N}]$ is isomorphic to ℓ_2 in $L_p[0,1]$ [$(r_n)_{n \in \mathbb{N}}$ are the Rademacher functions on $[0,1]$ (see IV.2.11)]. Then apply Corollary IV.3.8 to get an isometric copy of ℓ_2 in $L_p[0,1]$.

Theorem IV.4.2. For all $p \in [1, +\infty[$, ℓ_p is isometric to a complemented subspace of $L_p[0,1]$.

Proof. We already saw in Corollary IV.1.6 how to find ℓ_p in $L_p[0,1]$. Let us recall the construction:

Let $(I_n)_{n \in \mathbb{N}}$ be a sequence of intervals of $[0,1]$ such that

$$I_n \cap I_m = \varnothing \quad \text{if } m \neq n$$
$$[0,1] = \bigcup_{n=0}^{\infty} I_n$$

Define, for all n, $f_n = 1_{I_n}/\lambda(I_n)^{1/p}$. Then $(f_n)_{n \in \mathbb{N}}$ is a *peak sequence* in the sense of Definition IV.2.10. Moreover we can write, for all $(a_i)_{i \in \mathbb{N}} \in \mathbb{R}^{(\mathbb{N})}$,

$$\left\| \sum_{i=1}^{\infty} a_i f_i \right\|_p^p = \sum_{i=1}^{\infty} |a_i|^p \| f_n \|_p^p = \sum_{i=1}^{\infty} |a_i|^p$$

Thus $\overline{Span} \ [f_n, \ n \in \mathbb{N}]$ is isometric to ℓ_p.

Let P be defined by

$$P : L_p[0,1] \rightarrow \overline{Span} \ [f_n, \ n \in \mathbb{N}]$$

$$f \mapsto \sum_{n=0}^{\infty} \frac{1_{I_n}}{\lambda(I_n)^{1/p}} \int_{I_n} f \ d\lambda$$

Then P is a projection of norm 1 onto $\overline{Span} \ [f_n, \ n \in \mathbb{N}]$.

To determine the other values of q such that ℓ_q embeds in $L_p[0,1]$, we have to distinguish two cases, depending on the position of p in relation to 2.

In the case $p \geq 2$, p and 2 are the only values of q such that ℓ_q embeds in $L_p[0,1]$ because of Kadec-Pelcynski's theorem [KP]:

Theorem IV.4.3. Let $2 \leq p < +\infty$ and $(x_n)_{n \in \mathbb{N}}$ be a bounded sequence in $L_p[0,1]$ such that $(x_n)_{n \in \mathbb{N}}$ converges to 0 for $\sigma(L_p, L_{p'})$ and not in norm. Then, either $(x_n)_{n \in \mathbb{N}}$ has a subsequence which is equivalent to the unit vector basis of ℓ_2 or for all $\epsilon > 0$ it has a subsequence which is $(1 + \epsilon)$-equivalent to the unit vector basis of ℓ_p.

Before proving this theorem, let us state two obvious corollaries, which are other ways to formulate Theorem IV.4.3. These results can be compared with Proposition I.5.4 for the corresponding results on ℓ_p.

Corollary IV.4.4. Let $2 \leq p < +\infty$. Then every infinite-dimensional subspace of $L_p[0,1]$ either is isomorphic to ℓ_2 or for all $\epsilon > 0$ it contains a subspace which is $(1 + \epsilon)$-isomorphic to ℓ_p.

Corollary IV.4.5. Let $2 \leq p < +\infty$. If X is a subspace of $L_p[0,1]$ which is isomorphic to ℓ_q, then $q = 2$ or $q = p$.

Proof of Theorem IV.4.3. Fix $p \geq 2$. By taking a subsequence of $(x_n)_{n \in \mathbb{N}}$, if necessary, we can suppose that $(\| x_n \|)_{n \in \mathbb{N}}$ is "almost" constant. By scaling, it is thus enough to prove this result with a sequence $(x_n)_{n \in \mathbb{N}}$ such that $\| x_n \| = 1$ for all $n \in \mathbb{N}$. Define, for all $\epsilon > 0$,

$$A(\epsilon,p) = \{x \in L_p[0,1] / \lambda\{t \in [0,1] / | x(t) | \geq \epsilon \| x \|_p\} \geq \epsilon\}$$

Lemma IV.4.6. For all $r \in [1,p[$ and $x \in A(\epsilon,p)$, we have

$$\epsilon^{1 + 1/r} \| x \|_p \leq \| x \|_r \leq \| x \|_p$$

Proof of Lemma IV.4.6. Let $P_\epsilon = \{t \in [0,1] / | x(t) | \geq \epsilon \| x \|_p\}$. Then we can write

$$\| x \|_r = \left(\int_0^1 | x(t) |^r \, dt \right)^{1/r} \geq \left(\int_{P_\epsilon} | x(t) |^r \, dt \right)^{1/r}$$

$$\geq \epsilon \| x \|_p \, \lambda(P_\epsilon)^{1/r} \geq \epsilon^{1 + 1/r} \| x \|_p$$

This proves this lemma.

To prove Theorem IV.4.3, we have to distinguish between two cases:

First case: There exists $\epsilon > 0$ such that $x_n \in A(\epsilon,p)$ for all $n \in \mathbb{N}$. In that case, we are going to show that there exists a subsequence $(x_{n_k})_{k \in \mathbb{N}}$ of $(x_n)_{n \in \mathbb{N}}$ which is equivalent to the unit vector basis of ℓ_2.

Indeed, by Theorems I.3.12 and I.1.17, we can extract a basic unconditional subsequence $(x_{n_k})_{k\in\mathbb{N}}$ of $(x_n)_{n\in\mathbb{N}}$. This implies in particular

$$\left\|\sum_{i=0}^{\infty} a_i x_{n_i}\right\|_p \sim \left(\int_0^1 \left\|\sum_{i=0}^{\infty} r_i(t) a_i x_{n_i}\right\|_p^p dt\right)^{1/p}$$

Then using the fact that L_p is of Rademacher type 2 (Theorem IV.2.15), we get, for all $(a_i)_{i\in\mathbb{N}} \in \mathbb{R}^{(\mathbb{N})}$,

$$\left\|\sum_{i=0}^{\infty} a_i x_{n_i}\right\|_p \sim \left(\int_0^1 \left\|\sum_{i=0}^{\infty} r_i(t) a_i x_{n_i}\right\|_p^p dt\right)^{1/p}$$

$$\leq T_2 \left(\sum_{i=0}^{\infty} |a_i|^2 \|x_{n_i}\|^2\right)^{1/2} = T_2 \left(\sum_{i=0}^{\infty} |a_i|^2\right)^{1/2}$$

where T_2 is the Rademacher type 2 constant of L_p.

Using first the obvious fact that the Rademacher functions $(r_n)_{n\in\mathbb{N}}$ are orthogonal in $L_2[0,1]$, we get, for all $(a_i)_{i\in\mathbb{N}} \subset \mathbb{R}^{(\infty)}$,

$$\left\|\sum_{i=0}^{\infty} a_i x_{n_i}\right\|_p \sim \left(\int_0^1 \left\|\sum_{i=0}^{\infty} r_i(t) a_i x_{n_i}\right\|_p^2 dt\right)^{1/2}$$

$$\geq \left(\int_0^1 \int_0^1 \left|\sum_{i=0}^{\infty} r_i(t) a_i x_{n_i}(s)\right|^2 dt\, ds\right)^{1/2}$$

$$= \left(\sum_{i=0}^{\infty} |a_i|^2 \|x_{n_i}\|_2^2\right)^{1/2} \geq \left(\sum_{i=0}^{\infty} |a_i|^2\right)^{1/2}$$

So $(x_{n_k})_{k\in\mathbb{N}}$ is equivalent to the unit vector basis of ℓ_2.

Second case: For all $\epsilon > 0$, there exists $n \in \mathbb{N}$ such that $x_n \notin A(\epsilon,p)$. Let us first state some properties of $A(\epsilon,p)$:

(a) $\epsilon_2 < \epsilon_1 \Rightarrow A(\epsilon_2,p) \supset A(\epsilon_1,p)$
(b) $L_p[0,1] = \bigcup_{0<\epsilon<1} A(\epsilon,p)$
(c) $x \notin A(\epsilon,p) \Rightarrow \exists P \in \mathcal{B}$ such that

$$\begin{cases} \lambda(P) < \epsilon \\ \int_P |x(t)|^p\, dt \geq (1 - \epsilon^p)\|x\|_p^p \end{cases}$$

(a) is obvious. To prove (b), for $x \neq 0$, $x \in L_p[0,1]$, we consider

$$F(\epsilon) = \lambda\{t \in [0,1] / \frac{|x(t)|}{\|x\|_p} \geq \epsilon\}$$

F is left-continuous and decreases from 1 to 0 when ϵ increases from 0 to 1. So there is $\epsilon_0 > 0$ such that $F(\epsilon_0) \geq \epsilon_0$ and $x \in A(\epsilon_0, p)$.

To prove (c), suppose that $x \neq A(\epsilon, p)$; that is,

$$\lambda\{t \in [0,1]/|x(t)| \geq \epsilon \|x\|_p\} \leq \epsilon$$

and set as before:

$$P_\epsilon = \{t \in [0,1]/|x(t)| \geq \epsilon \|x\|_p\}$$

Then

$$\int_{P_\epsilon^c} |x(t)|^p \, dt \leq \epsilon^p \|x\|_p^p$$

and this proves that P_ϵ has the desired property.

We can now give the result in the case $p \geq 2$:

Lemma IV.4.7. Let $(x_n)_{n \in \mathbb{N}}$ be a normalized sequence in $L_p[0,1]$ such that for all $\epsilon > 0$, there exists $n_\epsilon \in \mathbb{N}$ such that $x_{n_\epsilon} \notin A(\epsilon, p)$. Then, for all $\delta > 0$, there exists a subsequence $(x_{n_k})_{k \in \mathbb{N}}$ of $(x_n)_{n \in \mathbb{N}}$ which is $(1 + \delta)$-equivalent to the unit vector basis of ℓ_p.

Proof of Lemma IV.4.7. We are going to construct $(x_{n_k})_{k \in \mathbb{N}}$ by induction: Take $\epsilon_0 = \frac{1}{2}$. By hypothesis, there exists $x_{n_0} \notin A(\epsilon_0, p)$. By (c), there exists $P_0 \in \mathcal{B}$ such that

$$\lambda(P_0) < \epsilon_0$$
$$\int_{P_0} |x_{n_0}(t)|^p \, dt \geq (1 - \epsilon_0^p)$$

Take $\epsilon_1 < 1/2^2$ and small enough to ensure that

$$\forall A \in \mathcal{B}, \qquad \lambda(A) < \epsilon_1 \Rightarrow \int_A |x_{n_0}(t)|^p \, dt < \epsilon_0^p$$

By hypothesis, there exists $x_{n_1} \notin A(\epsilon_1, p)$. By (c), there exists $P_1 \in \mathcal{B}$ such that

$$\lambda(P_1) < \epsilon_1$$
$$\int_{P_1} |x_{n_1}|^p \, dt \geq (1 - \epsilon_1^p)$$

Going on like this, we get sequences $(P_i)_{i \in \mathbb{N}}$ in $\mathcal{B}$, $(\epsilon_i)_{i \in \mathbb{N}}$ in $\mathbb{R}^+$, and

$(x_{n_i})_{i \in \mathbb{N}}$ in $L_p[0,1]$ such that:

(i) $\forall i \in \mathbb{N}, \qquad \lambda(P_i) < \epsilon_i < \dfrac{1}{2^{i+1}}$

(ii) $\forall i \in \mathbb{N}, \qquad \displaystyle\int_{P_i} |\, x_{n_i}(t)|^p \, dt \geq (1 - \epsilon_i^p)$

(iii) $\forall i \in \mathbb{N}, \qquad \displaystyle\int_{P_i} \sum_{j=0}^{i-1} |\, x_{n_j}(t)|^p \, dt \leq \epsilon_{i-1}^p$

We define, for all $i \in \mathbb{N}$,

$$P_i' = P_i \backslash (\bigcup_{j>i} P_j)$$

The sets $(P_i')_{i \in \mathbb{N}}$ are disjoint. For all $i \in \mathbb{N}$, we define

$$x_i' = \frac{x_{n_i} 1_{P_i'}}{\| \, x_{n_i} 1_{P_i'}\|_p}$$

Then $(x_i')_{i \in \mathbb{N}}$ is a sequence of normalized elements of $L_p[0,1]$ with disjoint supports. Thus $(x_i')_{i \in \mathbb{N}}$ is 1-equivalent to the unit vector basis of ℓ_p (see the proof of Theorem IV.4.2).

On the other hand, we have, for all $i \in \mathbb{N}$,

$$1 \geq \int_{P_i'} |\, x_{n_i}(t)|^p \, dt \geq \int_{P_i} |\, x_{n_i}(t)|^p \, dt - \sum_{j>i} \int_{P_j} |\, x_{n_i}(t)|^p \, dt$$

$$\geq (1 - \epsilon_i^p) - (\sum_{j>i} \epsilon_{j-1}^p)$$

$$\geq \left(1 - \frac{1}{2^{p(i+1)}}\right) - \left(\sum_{j>i} \frac{1}{2^{pj}}\right)$$

$$\geq 1 - \frac{1}{2^{p(i-1)}}$$

Then we can write

$$\| \, x_i' - x_{n_i} \|_p \leq \| \, x_i' - x_{n_i} 1_{P_i'} \|_p + \| \, x_{n_i} 1_{P_i'} - x_{n_i} \|_p$$

$$\leq \left\| x_{n_i} 1_{P_i'} \left(1 - \frac{1}{\| \, x_{n_i} 1_{P_i'} \|}\right) \right\| + \left(\int_{P_i'^c} |\, x_{n_i}(t)|^p \, dt\right)^{1/p}$$

$$\leq \left(\frac{1}{\| \, x_{n_i} 1_{P_i'} \|} - 1\right) + \left(\frac{1}{2^{p(i-1)}}\right)^{1/p}$$

$$\leq \left(\frac{1}{2^{p(i-1)} - 1}\right)^{1/p} + \frac{1}{2^{p(i-1)}}$$

$$\leq \frac{1}{2^{i-2}} + \frac{1}{2^{p(i-1)}} \leq \frac{1}{2^{i-3}}$$

So, for all $\delta > 0$, by Proposition I.1.15, $(x_{n_i})_{i \in \mathbb{N}}$ has a subsequence which is $(1 + \delta)$-equivalent to a subsequence of $(x_i')_{i \in \mathbb{N}}$. Hence this subsequence of $(x_n)_{n \in \mathbb{N}}$ is $(1 + \delta)$-equivalent to the unit vector basis of ℓ_p.

This finishes the proof of Lemma IV.4.7 and thus of Theorem IV.4.3.

Remark. Lemma IV.4.7 is also true for $1 \leq p < 2$. However, as we will see below, Theorem IV.4.3 is not true for these values of p.

In the case $1 \leq p < 2$, there are other values of q such that ℓ_q embeds in L_p: The next result is well known; it is a direct consequence of the existence of *q-stable random variables* in $L_p[0,1]$:

Theorem IV.4.8. Let $1 \leq p < 2$. Then

1. If $1 \leq q < p$ or if $q > 2$, ℓ_q is not isomorphic to any subspace of $L_p[0,1]$.
2. For all $q \in]p,2[$, ℓ_q is isometric to a subspace of $L_p[0,1]$.

Proof. 1 is a consequence of Theorem IV.2.15: $L_p[0,1]$ is of Rademacher type p, so it cannot contain ℓ_q for $q < p$ which is not of Rademacher type p. $L_p[0,1]$ is of Rademacher cotype 2, so it cannot contain ℓ_q for $q > 2$ which is not of Rademacher cotype 2.

To prove 2 we need to define *q-stable random variables* on [0,1] for $0 < q < 2$ (see [Fe]).

Definition IV.4.9. For $0 < q < 2$, X_q is called a *q-stable standard random variable* on [0,1] if $\hat{X}_q(t) = e^{-|t|^q}$, where $\hat{X}_q$ denotes the Fourier transform of X_q; that is,

$$\hat{X}_q(t) = \int_0^1 e^{itX_q(\omega)} \, d\omega$$

Existence of q-stable random variables, $0 < q < 2$. Let X be a symmetric r.v. on [0,1] such that

$$\lambda\{\omega \in [0,1]/|X(\omega)| > a\} = \frac{1}{a^q} \wedge 1$$

The density f_X of X on $\mathbb{R}$ is defined by

$$\begin{cases} f_X(x) = 0, & |x| \le 1 \\ f_X(x) = \dfrac{q}{2|x|^{q+1}}, & |x| > 1 \end{cases}$$

Then the Fourier transform $\hat{X}$ of X is the following:

$$\hat{X}(t) = \int_1^{+\infty} e^{itx} \frac{q}{2x^{q+1}} dx + \int_{-\infty}^{-1} e^{itx} \frac{q}{2(-x)^{q+1}} dx$$

$$= q \int_1^{+\infty} \frac{e^{itx} + e^{-itx}}{2} \frac{dx}{x^{q+1}} = q \int_1^{+\infty} \cos tx \frac{dx}{x^{q+1}}$$

Let $(X_n)_{n \in \mathbb{N}}$ be a sequence of independent, identically distributed random variables (i.i.d. r.v.) with the distribution above.
We consider the Fourier transform F_N of the r.v. $\sum_{i=0}^{N-1} X_i/N^{1/q}$. We are going to prove that for all $t \in \mathbb{R}$, $(F_N(t))_{N \in \mathbb{N}}$ converges to $e^{-qK_q|t|^q}$ where

$$K_q = \int_0^{+\infty} (1 - \cos x) \frac{dx}{x^{q+1}}$$

(Note that K_q is finite because $0 < q < 2$ and $1 - \cos t \le t^2/2$.)
We can write

$$F_N(t) = \left(\frac{\sum_{i=0}^{N-1} X_i}{N^{1/q}}\right)^{\wedge}(t) = \int_0^1 \exp\left(\frac{it \sum_{i=0}^{N-1} X_i(\omega)}{N^{1/q}}\right) d\omega$$

$$= \prod_{i=0}^{N-1} \hat{X}_i\left(\frac{t}{N^{1/q}}\right) \quad \text{(by independence)}$$

$$= \left[q \int_1^{+\infty} \cos\left(\frac{tx}{N^{1/q}}\right) \frac{dx}{x^{q+1}}\right]^N$$

Thus

$$\text{Log } F_N(t) = N \log\left[q \int_1^{+\infty} \cos\left(\frac{t}{N^{1/q}} x\right) \frac{dx}{x^{q+1}}\right]$$

$$= N \log \left[\frac{q}{N} \int_{N^{-1/q}}^{+\infty} \cos tx \, \frac{dx}{x^{q+1}} \right]$$

$$= N \log \left[-\frac{q}{N} \left(-\int_{N^{-1/q}}^{+\infty} \frac{dx}{x^{q+1}} \right. \right.$$

$$\left. \left. -\int_{N^{-1/q}}^{+\infty} (1 - \cos tx) \frac{dx}{x^{q+1}} \right) \right]$$

$$= N \log \left[1 - \frac{q}{N} \int_{N^{-1/q}}^{+\infty} (1 - \cos tx) \frac{dx}{x^{q+1}} \right]$$

When N tends to $+\infty$, this last expression is equivalent to

$$-q \int_{N^{-1/q}}^{+\infty} (1 - \cos tx) \frac{dx}{x^{q+1}}$$

which converges to

$$-q \int_0^{+\infty} (1 - \cos tx) \frac{dx}{x^{q+1}} = - q K_q |t|^q$$

Hence $(F_N(t))_{N \in \mathbb{N}}$ converges to $e^{-qK_q|t|^q}$ when N tends to $+\infty$. We now apply Paul Levy's theorem [Fe].

If $(Y_n)_{n \in \mathbb{N}}$ is a sequence of r.v. such that $(\hat{Y}_n(t))_{n \in \mathbb{N}}$ converges to $\varphi(t)$ for all $t \in \mathbb{R}$ and φ is continuous on $\mathbb{R}$, then there exists an r.v. Y such that $\hat{Y}(t) = \varphi(t)$.

This proves that there exists an r.v. X_q' such that $\hat{X}_q'(t) = e^{-qK_q|t|^q}$. Then $X_q = (1/(qK_q)^{1/q}) X_q'$ is a standard q-stable random variable.

The proof of Theorem IV.4.8 is a direct consequence of the following properties of q-stable r.v.:

Proposition IV.4.10. If $0 < q < 2$, let X_q be a standard q-stable random variable. Then

(i) $X_q \notin L_q[0,1]$.

(ii) $X_q \in L_p[0,1], \forall p \in]0,q[$.

(iii) If $(X_n)_{n \in \mathbb{N}}$ is a sequence of i.i.d. standard q-stable random variables, then for all $p < q$, $\overline{\text{Span}} [X_n, n \in \mathbb{N}]$ is isometric to ℓ_q in $L_p[0,1]$.

Proof. To prove (i) and (iii), we compute $\int_0^{+\infty} [1 - \hat{X}_q(t)]\, dt/t^{p+1}$ for $0 < p < q$ by two different methods:

(1)
$$\int_0^{+\infty} (1 - \hat{X}_q(t)) \frac{dt}{t^{p+1}} = \int_0^{+\infty} (1 - e^{-|t|^q}) \frac{dt}{t^{p+1}}$$

This last integrand is finite for $0 < p < q$ and tends to $+\infty$ when p tends to q.

(2)
$$\int_0^{+\infty} (1 - \hat{X}_q(t)) \frac{dt}{t^{p+1}} = \int_0^{+\infty} \int_0^1 (1 - e^{itX_q(\omega)})\, d\omega\, \frac{dt}{t^{p+1}}$$

$$= \int_0^1 \int_0^{+\infty} (1 - \cos tX_q(\omega)) \frac{dt}{t^{p+1}}\, d\omega$$

(because X_q is symmetric)

$$= \int_0^1 |X_q(\omega)|^p\, d\omega \int_0^{+\infty} (1 - \cos t) \frac{dt}{t^{p+1}}$$

$$= \| X_q \|_p \int_0^{+\infty} (1 - \cos t) \frac{dt}{t^{p+1}}$$

Since $0 < p < 2$, $\int_0^{+\infty} (1 - \cos t)\, dt/t^{p+1}$ is finite.

These two computations show that if $p < q$, $\| x_q \|_p$ is finite and tends to $+\infty$ when p tends to q, which proves (i) and (ii).

To prove (iii), it is enough to remark that for all $(x_i)_{i\in\mathbb{N}} \subset \mathbb{R}^{(\mathbb{N})}$:

$$\left(\sum_{i=0}^{\infty} a_i X_i \right)^{\wedge} (t) = \prod_{i=0}^{\infty} e^{-|a_i|^q|t|^q} = e^{-\sum_{i=0}^{\infty} |a_i|^q|t|^q}$$

Thus, the distribution of $\sum_{i=0}^{\infty} a_i X_i$ is the same as the distribution of $(\sum_{i=0}^{\infty} |a_i|^q)^{1/q} X_0$, and so, for $1 \leq p < q$,

$$\left\| \sum_{i=0}^{\infty} a_i X_i \right\|_p = \left(\sum_{i=0}^{\infty} |a_i|^q \right)^{1/q} \| X_0 \|_p$$

which proves (iii).

Corollary IV.4.11. If $1 \leq p \leq q \leq 2$, then $L_q[0,1]$ is isometric to a subspace of $L_p[0,1]$.

Proof. By Theorems IV.4.1, IV.4.2, and IV.4.8 we know that for $1 \leq p \leq q \leq 2$, ℓ_q is isometric to a subspace of $L_p[0,1]$. Thus, any separable subspace of an ultrapower of ℓ_q is isometric to a separable subspace of an ultrapower of $L_p[0,1]$; it is also isometric to a subspace of $L_p[0,1]$ by Corollary IV.1.6 and Theorem IV.3.1.

So to prove Corollary IV.4.11, it is enough to show that $L_q[0,1]$ is isometric to a subspace of any ultrapower of ℓ_q or, equivalently, that $L_q[0,1]$ is f.r. in ℓ_q (see Proposition II.1.7.3 and Theorem II.1.8).

Let $Y = \overline{\text{Span}} [f_0, \ldots, f_n]$ be a finite-dimensional subspace of $L_p[0,1]$. Fix $\epsilon > 0$. Then, by the density of simple functions in $L_p[0,1]$, there exist $(p_i)_{0 \leq i \leq n}$ in $\mathbb{N}$, $(A_i^j)_{0 \leq i \leq n, 0 \leq j \leq p_i}$ in $\mathcal{B}$, and $(a_i^j)_{0 \leq i \leq n, 0 \leq j \leq p_i}$ in $\mathbb{R}$ such that

$$\forall i = 0, \ldots, n, \qquad \left\| f_i - \sum_{j=0}^{p_i} a_i^j 1_{A_i^j} \right\|_q \leq \epsilon$$

Let $(B_k)_{0 \leq k \leq K}$ be the thinnest partition of $\cup_{0 \leq i \leq n, 0 \leq j \leq p_i} A_i^j$, constructed from all intersections of the $(A_i^j)_{0 \leq i \leq n, 1 \leq j \leq p_i}$. Then the closed linear span Z of $(1_{B_k})_{0 \leq k \leq K}$ is isometric to a subspace of ℓ_q^{K+1}.

On the other hand, Y is at distance less than $(1 + \epsilon)$ from a subspace of Z. Thus Y is at distance less than $(1 + \epsilon)$ from a subspace of ℓ_q, and this proves Corollary IV.4.11.

5. ℓ_q's IN SUBSPACES OF $L_p[0,1]$

In this section we are going to characterize the set of q's such that ℓ_q embeds in a given subspace X of $L_p[0,1]$.

In the case $p \geq 2$, Corollaries IV.4.4 and IV.4.5 give a complete characterization of the set of q's such that X contains ℓ_q: it can be $\{2,p\}$ or $\{2\}$ or $\{p\}$.

In the case $1 \leq p < 2$, the situation is very different and more complicated. Moreover, the case of $L_1[0,1]$ is special, in particular because it is not reflexive. Let us give a first result in this space, which is due to M. Kadec and A. Pelcynski [KP]. It can also be found in [Be3] and [P1].

Theorem IV.5.1. Let X be a subspace of $L_1[0,1]$. Then either X is reflexive or, for every $\epsilon > 0$, X contains a complemented subspace which is $(1 + \epsilon)$-isomorphic to ℓ_1.

Proof. Since L_1 is stable, we already know by Corollaries III.5.2 and III.4.10 that either X is reflexive or it contains a $(1 + \epsilon)$-copy of ℓ_1.

Let us give a direct and easy proof of this somewhat stronger result. Let S_X be the unit sphere of X.

Suppose S_X is 1-equi-integrable. Then $\overline{B}_X$, the closed unit ball of X, is obviously also 1-equi-integrable. By Corollary IV.2.7, $\overline{B}_X$ is $\sigma(L_1,L_\infty)$-compact and thus $\sigma(X,X^*)$-compact. Hence X is reflexive.

Otherwise S_X is not 1-equi-integrable. Then there exists $\delta > 0$ such that

$$\lim_{a \to +\infty} \, \text{Sup}_{f \in S_X} \left(\int_{\{|f|>a\}} |f(t)| \, dt \right) = \delta > 0$$

This implies that there exist $(a_n)_{n \in \mathbb{N}} \subset \mathbb{R}$, $a_n \xrightarrow[n \to +\infty]{} +\infty$, and $(f_n)_{n \in \mathbb{N}}$ $\subset S_X$ such that

$$\forall n \in \mathbb{N}, \qquad \delta\left(1 - \frac{1}{n}\right) \le \int_{\{|f_n|>a_n\}} |f_n(t)| \, dt \le \delta\left(1 + \frac{1}{n}\right)$$

For all $n \in \mathbb{N}$, set $g_n = f_n \, 1_{\{|f_n|>a_n\}}$.

Then, for all $\epsilon > 0$, we can write

$$\lambda\{t \in [0,1]/|g_n(t)| > \epsilon \, \|g_n\|_1\} \le \lambda\{t \in [0,1]/|g_n(t)| > 0\}$$

$$\le \lambda\{t \in [0,1]/|f_n(t)| > a_n\} \le \frac{1}{a_n}$$

(by Tchebychev's inequality; see Proposition IV.2.3).

Hence $\lambda\{t \in [0,1]/|g_n(t)| > \epsilon \, \|g_n\|_1\}$ tends to 0 when n tends to $+\infty$. Thus, there exists $n \in \mathbb{N}$, such that g_n does not belong to the set $A(\epsilon,1)$, which was defined in Theorem IV.4.3. Since this can be done for all $\epsilon > 0$, it is then enough to apply Lemma IV.4.7 to prove Theorem IV.5.1.

Let us come back to the general case of L_p-spaces, for $1 \le p \le 2$. We need a definition:

Definition IV.5.2. Let X be a Banach space. We define (see [MP])

$$p(X) = \text{Sup} \, \{p \in [0,1]/X \text{ is of Rademacher type } p\}$$

$$q(X) = \text{Inf} \, \{q \in [1,+\infty]/X \text{ is of Rademacher cotype } q\}$$

We need a result of B. Maurey and G. Pisier [MP] which was improved by J. L. Krivine [K2]:

Theorem IV.5.3. For all Banach spaces X, $\ell_{p(X)}$ and $\ell_{q(X)}$ are f.r. in X.

The proof of this difficult theorem is given from different points of view in [MiS], [MiSh], and [P9].

Corollary IV.5.4. Let X be a Banach space. Then for all $p \in [p(X),2]$ $\cup \{q(X)\}$, ℓ_p is f.r. in X.

Proof. We know by Theorem IV.5.3 that $\ell_{p(X)}$ and $\ell_{q(X)}$ are f.r. in X. In particular, by the results of Section 3, $L_{p(X)}[0,1]$ is isometric to a subspace of an ultrapower of X. By Theorem IV.4.8, for $p(X) \le p \le 2$, ℓ_p is also isometric to a subspace of an ultrapower of X, which means that ℓ_p is f.r. in X

If X is a subspace of $L_p[0,1]$, $1 \le p < 2$, then it is possible to obtain more, namely that $\ell_{p(X)}$ is isomorphic to a subspace of X. The following results are due to M. Levy and the author [GLe].

Theorem IV.5.5. Let X be a subspace of $L_p[0,1]$, $1 \le p < 2$. Then for all $\epsilon > 0$, $\ell_{p(X)}$ is $(1 + \epsilon)$-isomorphic to a subspace of X.

Proof. Since $L_p[0,1]$ is isometric to a subspace of $L_1[0,1]$ by Corollary IV.4.11, it is sufficient to show this result for $p = 1$.
This theorem is a consequence of the following:

Theorem IV.5.6. Let X be a subspace of $L_1[0,1]$. Then if ℓ_p is f.r. in X, there exists $q \le p$ such that for all $\epsilon > 0$, ℓ_q is $(1 + \epsilon)$-isomorphic to a subspace of X.

Proof that Theorem IV.5.6 implies Theorem IV.5.5. By Theorem IV.5.3, $\ell_{p(X)}$ is f.r. in X, so by Theorem IV.5.6 there exists $q \le p(X)$ such that for all $\epsilon > 0$, ℓ_q is $(1 + \epsilon)$-isomorphic to a subspace of X. This number q cannot be strictly less than $p(X)$ because ℓ_q is not of Rademacher type more than q and X is, by definition of $p(X)$. So $q = p(X)$, and this proves Theorem IV.5.5.

Proof of Theorem IV.5.6. To prove this theorem, we are going to use all results in Chapters II and III. Let X be a subspace of $L_1[0,1]$ such that ℓ_p is f.r. in X for some $p \in [1,2]$. Then ℓ_p is a subspace of an ultrapower $Y = X^{\mathbb{N}}/\mathfrak{D}$ of X. Denote by Y_1 the closed linear span of X and ℓ_p in Y. Note that Y_1 is separable and stable.

Then, in the conic class of the type, generated by the unit vector basis of ℓ_p in Y_1, there exists an ℓ_q-type by Remark III.4.9. Using the arguments of Lemma III.5.3, it is easy to see that necessarily $q = p$. So there is an ℓ_p-type $\bar{\sigma}$ in $\mathcal{T}(Y_1)$.

Our aim is to prove that the conic class in $\mathcal{T}(X)$, generated by the restriction (in a sense to be defined) σ of $\bar{\sigma}$ to X, contains an ℓ_q-type for some $q \le p$.

For this we need four steps:

First step: Representation of $\| x - y \|_X$ and $\| \sigma * (-\tau) \|_X$: *magic formula.*

If we define K by $K = \int_0^{+\infty} (1 - \cos t) \, dt/t^2$, then K is positive and finite. Moreover if $u \in \mathbb{R}$, we have

$$K \, | u | = \int_0^{+\infty} (1 - \cos tu) \, \frac{dt}{t^2}$$

For $u, v \in \mathbb{R}^2$, we can write

$$1 - \cos t(u - v) = (1 - \cos tu) + (1 - \cos tv)$$
$$- (1 - \cos tu)(1 - \cos tu) - (\sin tu)(\sin tv)$$

Integrating this equality in t with respect to dt/t^2 on $[0, +\infty]$, we get

$$K(| u - v | - | u | - | v |)$$
$$= - \int_0^{+\infty} [(1 - \cos tu)(1 - \cos tv) + \sin tu \sin tv] \frac{dt}{t^2}$$

Applying this to $u = x(\omega)$ and $v = y(\omega)$, where $x, y \in X \subset L_1[0,1]$, and integrating in ω with respect to λ on $[0,1]$, we obtain

$$K(\| x - y \|_1 - \| x \|_1 - \| x \|_1) =$$
$$- \int_0^1 \int_0^{+\infty} [(1 - \cos tx(\omega))(1 - \cos ty(\omega))$$
$$+ \sin tx(\omega) \sin ty(\omega)] \frac{dt}{t^2} \, d\lambda(\omega)$$

Hence, if we denote by U the map

$$U : X \to \left[L_2\left([0,1] \times [0, +\infty[, \, d\lambda \otimes \frac{dt}{t^2} \right) \right]^2$$

$$x \mapsto U_x = (1 - \cos tx, \sin tx)$$

and by $\langle \, , \, \rangle$ the inner product in $[L_2([0,1] \times [0, +\infty[, \, d\lambda \otimes dt/t^2)]$ we get the *magic formula*:

$$K(\|x - y\|_x - \|x\|_x - \|y\|_x) = -\langle U_x, U_y \rangle$$

This expresses the fact that the norm of subspaces of $L_1[0,1]$ is of *negative type*, in the terminology of [Fe].

Let us denote by u and v the two coordinates of U, that is, $U = (u,v)$, with $u_x = 1 - \cos tx$ and $v_x = \sin tx$, and by H the closed linear span of the set $\{U_x, x \in X\}$ in $[L_2([0,1] \times [0, +\infty[, \, d\lambda \otimes dt/t^2)]^2$.

The next result gives first properties of the map U:

Proposition IV.5.7.

1. $\forall x \in X, \, 2K \|x\|_x = \|U_x\|_H^2$.
2. $\forall x,y \in X, \, 2K \|x - y\|_x = \|U_x - U_y\|_H^2$.
3. $\forall x,y \in X$,

$$1 - u_{x+y} = (1 - u_x)(1 - u_y) + v_x v_y$$

$$v_{x+y} = (1 - u_x)v_y + (1 - u_y)v_x$$

$$u_{-x} = u_x \quad \text{and} \quad v_{-x} = -v_x$$

$\forall t, \lambda \in \mathbb{R}$,

$$u_{\lambda x}(t) = u_x(\lambda t) \quad \text{and} \quad v_{\lambda x}(t) = v_x(\lambda t)$$

4. For all $\sigma \in \mathcal{T}(X)$ there exists $U_\sigma \in H$ such that σ is represented by the bounded sequence $(x_n)_{n \in \mathbb{N}}$ and the ultrafilter $\mathcal{U}$ in X if and only if U_σ is the $\sigma(H, H^*)$-limit of $(U_{x_n})_{n \in \mathbb{N}}$ in H, along $\mathcal{U}$.

Proof

1. Take $x = y$ in the magic formula.
2. In the magic formula, replace $K \|x\|_x$ and $K \|y\|_x$ by $\frac{1}{2} \|U_x\|_2^2$ and $\frac{1}{2} \|U_y\|_2^2$. We get

$$K \|x - y\|_x = \frac{1}{2} [\|U_x\|_H^2 + \|U_y\|_H^2 - 2\langle U_x, U_y \rangle]$$

$$= \frac{1}{2} \|U_x - U_y\|_H^2$$

3. This is an obvious consequence of trigonometric formulas if we write

$$u_x = 1 - \cos tx$$
$$v_x = \sin tx$$

4. Let σ be a type on X and $((x_n)_{n\in\mathbb{N}}, \mathcal{U})$ a representation of σ by a bounded sequence of X and a ultrafilter $\mathcal{U}$. By 1, $(U_{x_n})_{n\in\mathbb{N}}$ is a bounded sequence in H. Call U the $\sigma(H,H^*)$-limit of $(U_{x_n})_{n\in\mathbb{N}}$ along $\mathcal{U}$. Then, since we can write for all x in X

$$K\,[\| x_n - x \|_X - \| x \|_X - \| x_n \|_X] = - \langle U_{x_n}, U_x \rangle$$

we get, by taking limits of both members along $\mathcal{U}$, for all x in X

$$K\,[\sigma(-x) - \| x \|_X - \| \sigma \|_X] = - \langle U, U_x \rangle$$

Hence, if $((y_n)_{n\in\mathbb{N}}, \mathcal{V})$ is another representation of σ in X and U' is the $\sigma(H,H^*)$-limit of $(U_{y_n})_{n\in\mathbb{N}}$ along $\mathcal{V}$, we will get, for all x in X

$$\langle U, U_x \rangle = \langle U', U_x \rangle$$

Thus, for all h in H, $\langle U, h \rangle = \langle U', h \rangle$, and then this implies that $U = U'$.

So, U does not depend on the representation of σ. We can call it U_σ. The equivalence which is announced in 4 is an obvious consequence of the construction of U_σ.

The map U can be extended to $\mathcal{T}(X)$ by Proposition IV.5.7(4). We obtain a map, called U too, from $\mathcal{T}(X)$ onto H. This allows us to extend the magic formula to types on X:

Corollary IV.5.8. The magic formula is true for types on X, that is,

$$\forall \sigma, \tau \in \mathcal{T}(X), \qquad K(\| \sigma * (-\tau) \|_X - \| \sigma \|_X - \| \tau \|_X) = -\langle U_\sigma, U_\tau \rangle$$

Moreover:

1. $\forall \sigma \in \mathcal{T}(X), 2K \| \sigma \|_X \geq \| U_\sigma \|_H^2$.
2. $\forall \sigma, \tau \in \mathcal{T}(X), 2K \| \sigma * (-\tau) \|_X \geq U_\sigma - U_\tau \|_H^2$.
3. $\forall \sigma, \tau \in \mathcal{T}(X),$

$$1 - u_{\sigma*\tau} = (1 - u_\sigma)(1 - u_\tau) + v_\sigma v_\tau$$

$$v_{\sigma*\tau} = (1 - u_\sigma)v_\tau + (1 - u_\tau)v_\sigma$$

$$u_{-\sigma} = u_\sigma \quad \text{and} \quad v_{-\sigma} = -v_\sigma$$

$\forall t, \lambda \in \mathbb{R}$,

$$u_{\lambda\sigma}(t) = u_\sigma(\lambda t) \quad \text{and} \quad v_{\lambda\sigma}(t) = v_\sigma(\lambda t)$$

4. $(\sigma_n)_{n\in\mathbb{N}}$ converges to σ in $\mathcal{T}(X)$ if and only if $(U_{\sigma_n})_{n\in\mathbb{N}}$ converges to U_σ for $\sigma(H, H^*)$.

Proof. Let us define $\sigma = ((x_n)_{n\in\mathbb{N}}, \mathcal{U})$ and $\tau = ((y_m)_{n\in\mathbb{N}}, \mathcal{V})$. We can write

$$\lim_{n,\mathcal{U}} \lim_{m,\mathcal{V}} K \left(\| x_n - y_m \|_X - \| x_n \|_X - \| y_m \|_X \right) = \lim_{n,\mathcal{U}} \lim_{m,\mathcal{V}} - \langle U_{x_n}, U_{y_m} \rangle$$

That is,

$$K(\| \sigma * (-\tau) \|_X - \| \sigma \|_X - \| \tau \|_X) = -\langle U_\sigma, U_\tau \rangle$$

To prove 1, 2, 3, 4, we just take limits in an appropriate sense in analogous properties of Proposition IV.5.7. Let us show, for instance, the first property: Since $2K \| x_n \|_X = \| U_{x_n} \|_H^2$, we obtain

$$2K \| \sigma \| = 2K \lim_{n,\mathcal{U}} \| x_n \|_X$$

$$= \lim_{n,\mathcal{U}} \| U_{x_n} \|_H^2 \geq \| \sigma(H, H^*) - \lim_{n,\mathcal{U}} U_{x_n} \|_H^2 = \| U_\sigma \|_H^2$$

Remark IV.5.9. If we apply the magic formula to a type σ on X and a realized type χ_x on X, we get also, for all x in X and all σ in $\mathcal{T}(X)$,

$$K[\sigma(-x) - \| x \|_X - \| \sigma \|_X] = -\langle U_\sigma, U_x \rangle$$

Second step: Extensions and restrictions of types.

1. If $\sigma = ((x_n)_{n\in\mathbb{N}}, \mathcal{U})$ is a type on $X \subset L_1[0,1]$, we can *extend* it to Y_1; indeed, every $\tilde{y} \in Y_1$ is represented by a sequence $(y_n)_{n\in\mathbb{N}}$ in X and we can define

$$\tilde{\sigma}(\tilde{y}) = \lim_{k,\mathcal{D}} \sigma(y_k)$$

Then

$$\tilde{\sigma}(\tilde{y}) = \lim_{k,\mathcal{D}} \lim_{n,\mathcal{U}} \| x_n + y_k \|_X = \lim_{n,\mathcal{U}} \lim_{k,\mathcal{D}} \| x_n + y_k \|_X$$

$$= \lim_{n,\mathcal{U}} \| \tilde{y} + x_n \|_{Y_1}$$

(As usual, we have identified X to a subspace of Y_1.) This proves that this extension $\tilde{\sigma}$ is a type on Y_1.

2. On the other hand, if $\check{\sigma}$ is a type on Y_1, we can *restrict* it to X; indeed, if $\check{\sigma} = ((\check{x}_n)_{n\in\mathbb{N}},\mathfrak{U})$, we can state for all $x \in X$

$$\sigma(x) = \lim_{n,\mathfrak{U}} \| x + \check{x}_n \|_{Y_1}$$

If for all $n \in \mathbb{N}$, $\check{x}_n$ is represented by a sequence $(x_n^k)_{k\in\mathbb{N}}$ in X, this implies

$$\sigma(x) = \lim_{n,\mathfrak{U}} \lim_{k,\mathfrak{D}} \| x + x_n^k \|_X$$

By a diagonal process, it is obvious to see that this proves that σ is a type on X.

Remark IV.5.10

1. It is clear that if we extend a type σ on X to Y_1 and then restrict it to X, we find σ again. On the other hand, if we restrict a type $\check{\sigma}$ on Y_1 to X and then extend it to Y_1, we do not find $\check{\sigma}$ again in general.
2. Extension or restriction does not change the norm of types.
3. Since Y_1 is a subspace of an ultrapower of X, the magic formula can be extended to Y_1: If $\check{x} = (x_k)_{k\in\mathbb{N}}$ and $\check{y} = (y_k)_{k\in\mathbb{N}}$ belong to Y_1, we can write

$$\lim_{k,\mathfrak{D}} K(\| x_k - y_k \|_X - \| x_k \|_X - \| y_k \|_X) = -\lim_{k,\mathfrak{D}} \langle U_{x_k},U_{y_k}\rangle$$

Let us denote by $\check{H}$ the ultrapower $H^{\mathbb{N}}/\mathfrak{D}$ of H. Then the inner product on $\check{H}$ is given by

$$\langle \check{h},\check{\ell}\rangle_{\check{H}} = \lim_{k,\mathfrak{D}} \langle h_k,\ell_k\rangle_H \quad \text{if} \begin{cases} \check{h} = (h_k)_{k\in\mathbb{N}} \\ \check{\ell} = (\ell_k)_{k\in\mathbb{N}} \end{cases}$$

Thus if we set $U_{\check{x}} = (U_{x_k})_{k\in\mathbb{N}}$ and $U_{\check{y}} = (U_{y_k})_{k\in\mathbb{N}}$ in $\check{H}$, we get

$$K(\| \check{x} - \check{y} \|_{Y_1} - \| \check{x} \|_{Y_1} - \| \check{y} \|_{Y_1}) = -\langle U_{\check{x}},U_{\check{y}}\rangle$$

Thus, the magic formula on X and the map U from X to H can be extended to Y_1, U taking its values in $\check{H}$ in this case. It is then clear that properties 1, 2, 3, 4 of Proposition IV.5.7 extend to Y_1 with no change.

4. As was done for Corollary IV.5.8, we can also extend the magic formula on Y_1 and the map $U: Y_1 \rightarrow \check{H}$ to $\mathcal{T}(Y_1)$, U still taking its

values in $\bar{H}$. Properties 1, 2, 3, 4 of Corollary IV.5.8 remain valid in $\mathcal{T}(Y_1)$ and Remark IV.5.9 also.

By construction, we obtain:

Proposition IV.5.11

1. Let $\bar{\sigma}$ belong to $\mathcal{T}(Y_1)$ and let σ be its restriction to X. If $U_{\bar{\sigma}}$ is represented by $(h_k)_{k \in \mathbb{N}}$ in $\bar{H} = H^{\mathbb{N}}/\mathcal{D}$, then we have

$$U_\sigma = \sigma(H, H^*) - \lim_{k,\mathcal{D}} h_k$$

2. Let σ belong to $\mathcal{T}(X)$ and let $\bar{\sigma}$ be its extension to Y_1. Then $U_{\bar{\sigma}} = U_\sigma$ belongs to H.

Proof

1. The magic formula in Y_1 gives for all $\tilde{x}$ in Y_1

$$K(\bar{\sigma}(-\tilde{x}) - \|\tilde{x}\|_{Y_1} - \|\bar{\sigma}\|_{Y_1}) = -\langle U_{\bar{\sigma}}, U_{\tilde{x}} \rangle$$

Thus if $U_{\bar{\sigma}} = (h_n)_{n \in \mathbb{N}} \in \bar{H}$ and $x \in X$ we get

$$K(\sigma(-x) - \|x\|_X - \|\sigma\|_X) = -\lim_{k,\mathcal{D}} \langle h_k, U_x \rangle$$

Hence, for all x in X, $\langle U_\sigma, U_x \rangle = \lim_{k,\mathcal{D}} \langle h_k, U_x \rangle$, which proves 1.
2. Is true by definition, H being identified as usual to a subspace of $\bar{H}$.

Remarks

1. The relation between $U_{\bar{\sigma}}$ and U_σ given by Proposition IV.5.11.1 proves that it is false in general that the restriction of $\bar{\sigma} * \bar{\tau}$ to X is $\sigma * \tau$ because $\sigma(H, H^*)$-limits do not pass to products. On the contrary, the extension of $\sigma * \tau$ to Y_1 is $\bar{\sigma} * \bar{\tau}$ by 2.
2. Moreover, for all $\lambda \in \mathbb{R}$, obviously the restriction of $\lambda\bar{\sigma}$ to X is $\lambda\sigma$ and the extension of $\lambda\sigma$ to Y_1 is $\lambda\bar{\sigma}$.

Third step: Positive types on X.

Here we are going to consider only symmetric types on X. For a symmetric type σ, by Corollary IV.5.8.3, we have $v_\sigma = 0$.

Definition IV.5.12.   A symmetric type θ in $\mathcal{T}(X)$ is said to be *positive* if $1 - u_\theta \geq 0$.

Example.   If θ is an ℓ_q-type for some $q \in [1,2]$, θ is positive. Indeed,

$$1 - u_\theta = 1 - u_{2 - 1/q_\theta * 2 - 1/q_\theta} = (1 - u_{2 - 1/q_\theta})^2 \quad \text{and} \quad 1 - u_\theta \geq 0$$

The next result studies extensions, restrictions, and convolutions of positive types:

Proposition IV.5.13

1. If θ is a positive type on X, its extension $\tilde{\theta}$ to Y_1 is also positive.
2. If $\tilde{\theta}$ is a positive type on Y_1, its restriction θ to X is also positive.
3. If θ is a positive type, for all $(a_0, \ldots, a_k) \in \mathbb{R}^{k+1}$, $a_0\theta * \cdots * a_k\theta$ is also a positive type.

Proof.   1 and 2 are immediate consequences of Proposition IV.5.11. 3 is a consequence of formulas of Corollary IV.5.8.

The next lemma and its corollary are the key properties of positive types:

Lemma IV.5.14.   Let $\tilde{\sigma}$ belong to $\mathcal{S}(Y_1)$, and let σ be its restriction to X. Let θ be a positive type on X and let $\tilde{\theta}$ be its extension to Y_1. Then

$$\| \sigma * \sigma * \theta \|_X \geq \| \tilde{\sigma} * \tilde{\sigma} * \tilde{\theta} \|_{Y_1}$$

Proof.   Since we are working with symmetric types, we can identify maps U and u because v is null. Then we can write

$$K \| \sigma * \sigma * \theta \|_X = K(\| \sigma * \sigma \|_X + \| \theta \|_X) - \langle u_{\sigma * \sigma}, u_\theta \rangle_H$$

$$= K(2 \| \sigma \|_X + \| \theta \|_X) - \langle u_\sigma, u_\sigma \rangle_H - \langle 2u_\sigma - u_\sigma^2, u_\theta \rangle_H$$

$$= K(2 \| \sigma \|_X + \| \theta \|_X) - \langle u_\sigma^2, 1 - u_\theta \rangle_H - 2 \langle u_\sigma, u_\theta \rangle_H$$

Of course, by the same computation we get

$$K \| \tilde{\sigma} * \tilde{\sigma} * \tilde{\theta} \|_{Y_1} = K(2 \| \tilde{\sigma} \|_{Y_1} + \| \tilde{\theta} \|_{Y_1})$$

$$- \langle u_{\tilde{\sigma}}, 1 - u_{\tilde{\theta}} \rangle_{\tilde{H}} - 2 \langle u_{\tilde{\sigma}}, u_{\tilde{\theta}} \rangle_{\tilde{H}}$$

We know that $\| \tilde{\sigma} \|_{Y_1} = \| \sigma \|_X$ and $\| \tilde{\theta} \|_{Y_1} = \| \theta \|_X$. Moreover, because

u_θ is in H, $\langle u_\sigma, u_\theta \rangle_H = \langle u_{\tilde\sigma}, u_\theta \rangle_{\tilde H}$ by Proposition IV.5.11. Thus, we have just have to prove that $\langle u_\sigma^2, 1 - u_\theta \rangle_{\tilde H} \geq \langle u_\sigma^2, 1 - u_\theta \rangle_H$.

Set $u_{\tilde\sigma} = (h_k)_{k \in \mathbb{N}}$ in $\tilde H$. Since $1 - u_\theta \geq 0$, this function is a density on $[0,1] \times [0, +\infty[$, and the inequality above expresses that the limit of the norm of $(h_k)_{k \in \mathbb{N}}$ in $L_2([0,1] \times [0, +\infty[, (1 - u_\theta)\, d\lambda \otimes dt/t^2)$ along $\mathcal{D}$ is more than the norm of u_σ in the same space. This is true because u_σ is the $\sigma(H, H^*)$-limit of $(h_n)_{n \in \mathbb{N}}$.

This proves Lemma IV.5.14.

Corollary IV.5.15. Let $\tilde\sigma$ be an ℓ_p-type on Y_1 and σ its restriction to X. Let θ be a positive type on X. Then

$$\| \sigma * \sigma * \theta \|_X \geq \| 2^{1/p}\sigma * \theta \|_X$$

Proof. We apply Lemma IV.5.14:

$$\| \sigma * \sigma * \theta \|_X \geq \| \tilde\sigma * \tilde\sigma * \tilde\theta \|_{Y_1} = \| 2^{1/p}\tilde\sigma * \tilde\theta \|_{Y_1}$$

Since the restriction of $2^{1/p}\tilde\sigma$ to X is $2^{1/p}\sigma$ and $\tilde\theta$ is the extension to Y_1 of $\theta \in \mathcal{T}(X)$, we can write

$$K \| 2^{1/p}\tilde\sigma * \tilde\theta \|_{Y_1} = K(2^{1/p} \| \tilde\sigma \|_{Y_1} + \| \tilde\theta \|_{Y_1}) - \langle u_{2^{1/p}\tilde\sigma}, u_\theta \rangle_{\tilde H}$$

$$= K(2^{1/p} \| \sigma \|_X + \| \theta \|_X) - \langle u_{2^{1/p}\sigma}, u_\theta \rangle_H$$

by Proposition IV.5.11. Thus

$$K \| 2^{1/p} \tilde\sigma * \tilde\theta \|_{Y_1} = K \| 2^{1/p} \sigma * \theta \|_X$$

and this proves Corollary IV.5.15.

Fourth Step: Proof of Theorem IV.5.6.

Suppose that ℓ_p is f.r. in X. Then there exists an ℓ_p-type $\tilde\sigma$ on Y_1. Let σ be its restriction to X.

Lemma IV.5.16. For all types τ belonging to the conic class $K(\sigma)$, generated by σ, we have

$$\| \tau * \tau \|_X \geq 2^{1/p} \| \tau \|_X$$

Proof of Lemma IV.5.16. First of all, suppose that there exist a_0, . . . , $a_k \in \mathbb{R}^{k+1}$ such that $\tau = a_0\sigma * \cdots * a_k\sigma$. By Proposition IV.5.13,

τ is a positive type. Thus, applying Corollary IV.5.15 $(k + 1)$ times, we get

$$\|(a_0\sigma * \cdots * a_k\sigma) * (a_0\sigma * \cdots * a_k\sigma)\|$$

$$= \|(a_0\sigma * a_0\sigma) * \cdots * (a_k\sigma * a_k\sigma)\|$$

$$\geq \| 2^{1/p} a_0\sigma * (a_1\sigma * a_1\sigma) * \cdots * (a_k\sigma * a_k\sigma)\|$$

$$\vdots$$

$$\geq \| 2^{1/p}(a_0\sigma * \cdots * a_k\sigma)\|$$

$$= 2^{1/p} \| a_0\sigma * \cdots * a_k\sigma \|$$

Thus, in this case, we have

$$\| \tau * \tau \|_X \geq 2^{1/p} \| \tau \|_X$$

If τ is a pointwise limit of a sequence $(\tau_n)_{n\in\mathbb{N}}$ of the form $\tau_n = a_0^n \sigma * \cdots * a_{k_n}^n\sigma$, we can write by the magic formula

$$K \| \tau * \tau \| = 2K \| \tau \| - \| u_\tau \|^2$$

$$= 2K \lim_n \| \tau_n \| - \| \sigma(H,H^*) - \lim u_{\tau_n} \|^2$$

$$\geq \lim_n [2K \| \tau_n \| - \| u_{\tau_n} \|^2]$$

$$= \lim_n K \| \tau_n * \tau_n \| \geq \lim_n 2^{1/p} K \| \tau_n \| = 2^{1/p} K \| \tau \|$$

This finishes the proof of Lemma IV.5.16.

Let's come back to the proof of Theorem IV.5.6: Indeed, the conic class $K(\sigma)$ contains an ℓ_q-type τ_0 for some $q \in [1,2]$. So we can write

$$\| \tau_0 * \tau_0 \| = 2^{1/q} \| \tau_0 \| \geq 2^{1/p} \| \tau_0 \|$$

Thus $q \leq p$, and for all $\epsilon > 0$, X contains a subspace which is $(1 + \epsilon)$-isomorphic to ℓ_q by Remark III.4.3.

Remark IV.5.17. Theorem IV.5.5 and Theorem IV.5.6 are false if X is not a subspace of $L_1[0,1]$: the space $\sum_n \oplus_p \ell_q^n$ for $1 \leq q < p \leq 2$ is stable and does not contain ℓ_q, but ℓ_q is f.r. in it and its best Rademacher type is q.

NOTES AND REMARKS

1. Beside L_p-spaces, ultrapowers of certain spaces are known. Let us mention Orlicz spaces [DCK1], $L_p(L_q)$ [LR], $L_p(X)$ [HLR], and r.i. function spaces [HL]. Moreover, the noncommutative analog of ultrapowers of L_1, namely ultrapowers of preduals of von Neumann algebras, are described in [Gr] as preduals of von Neumann algebras.

2. We would like to mention a very important result of H. P. Rosenthal on reflexive subspaces of $L_1[0,1]$:

   Theorem [Ros 2]. Let X be a reflexive subspace of $L_1[0,1]$. Then there exist $p > 1$ such that X embeds in $L_p[0,1]$.

3. In [Gu4], the author extends the techniques of Section IV.5 to prove that every $\sigma(L_p, L_{p'})$-null sequence in L_p ($p \geq 2$) and every sequence in L_p ($1 \leq p < 2$), which is equivalent to the unit vector basis of ℓ_2, has an almost symmetric subsequence. In [Gu5] and [GR], it is proved that this result is false for $\sigma(L_p, L_{p'})$-null sequences of L_p ($1 \leq p < 2$) which are not equivalent to the unit vector basis of ℓ_2.

4. More generally, it is possible (see [Ra4]) to describe subspaces of the form $\ell_r(\ell_s)$ of L_p and $L_p(L_q)$; namely r and s can be computed from p and q.

EXERCISES

1. Prove that the Borel σ-field $\mathcal{B}$ on $[0,1]$ is a complete metric space equipped with the distance $d(A,B) = \lambda(A \Delta B)^{1/p}$ ($1 \leq p < +\infty$).

2. Show that the Rademacher functions span a subspace isometric to ℓ_1 in L_∞. Show that sequences $(\epsilon_1, \epsilon_2, \ldots, \epsilon_{2^k}, 0, 0 \ldots)$, where $\epsilon_i = \pm 1$, span $\ell_1^{2^k}$ in ℓ_∞.

3. Show that a sequence of identically distributed r.v. in L_p is p-uniformly integrable.

4. Prove that every normalized peak sequence in L_p ($1 \leq p < +\infty$) defines an ℓ_p-type.

5. Show that every type σ on L_p ($1 \leq p < +\infty$) can be written as σ

$= \sigma_1 * \sigma_2$ where σ_1 is defined by a p-equi-integrable sequence and σ_2 is an ℓ_p-type. (Use Theorem IV.2.8.)

6. Show that $\sum_n \oplus_q \ell_q^n$ is stable and not superstable.

7. What are ultrapowers of a Hilbert space?

8. Find a type σ on L_p $(1 \le p < +\infty)$ which is defined by two different sequences $(x_n)_{n \in \mathbb{N}}$ and $(y_n)_{n \in \mathbb{N}}$ and such that $\| x_{n_k} - y_{m_k} \| \not\rightarrow_{k \to +\infty}$ 0 for any subsequences $(x_{n_k})_{k \in \mathbb{N}}$ and $(y_{m_k})_{k \in \mathbb{N}}$.

9. Quotients of ℓ_p, $1 \le p < +\infty$ (see [Ra2]).

   (a) Let $(x_n)_{n \in \mathbb{N}}$ be a bounded sequence in ℓ_p $(1 \le p < +\infty)$ and $\mathcal{U}$ a nontrivial ultrafilter on $\mathbb{N}$.

      Show that if $(x_n)_{n \in \mathbb{N}}$ converges to 0 for $\sigma(\ell_p, \ell_{p'})$ (or $\sigma(\ell_1, c_0)$ if $p = 1$), we get

      $$\forall x \in \ell_p, \qquad \lim_{n, \mathcal{U}} [\| x + x_n \|^p - \| x_n \|^p - \| x \|^p] = 0$$

   (b) Deduce from this that if σ is a type on ℓ_p $(1 \le p < +\infty)$, there exist $\bar{x} \in \ell_p$ and $a \ge 0$ such that

      $$\forall x \in \ell_p, \qquad \sigma(x) = [\| x + \bar{x} \|^p + a^p]^{1/p}$$

   (c) Show that every symmetric type on ℓ_p $(1 \le p < +\infty)$ is an ℓ_p-type.

   (d) Let Y be a (closed) subspace of ℓ_p, $1 \le p < +\infty$. If $p = 1$ we suppose that Y is $\sigma(\ell_1, c_0)$-closed.

      Let $(\mathring{\xi}_n)_{n \in \mathbb{N}}$ be a bounded sequence in ℓ_p / Y.

      (i) Show that there exists $\mathring{\xi} \in \ell_p / Y$ such that for every bounded sequence $(x_n)_{n \in \mathbb{N}}$ in ℓ_p such that $\forall n \in \mathbb{N}$, $x_n \in \mathring{\xi}_n$ there exists $x \in \mathring{\xi}$ such that

      $$\lim_{n, \mathcal{U}} x_n = x \quad \text{for } \sigma(\ell_p, \ell_{p'}) \text{ if } p \ne 1$$

      $$\text{for } \sigma(\ell_1, c_0) \text{ if } p = 1$$

      (ii) Prove that

      $$\forall \mathring{\phi} \in \ell_{p/Y},$$

      $$\lim_{n, \mathcal{U}} (\| \mathring{\xi}_n + \mathring{\phi} \|^p - \| \mathring{\xi}_n - \mathring{\xi} \|^p - \| \mathring{\xi} - \mathring{\phi} \|^p) = 0$$

      (iii) Deduce from this that ℓ_p / Y is stable and that every symmetric type on ℓ_p / Y is an ℓ_p-type.

(e) What can be said of quotients of ℓ_1 by non-$\sigma(\ell_1,c_0)$-closed subspaces?

10. Show that every reflexive subspace of L_1 is superreflexive (use Theorems IV.3.1, IV.5.3, IV.5.4, IV.5.5).

11. Let X be a Banach space which does not contain a subspace iso-morphic to c_0. Suppose that X has an unconditional and 1-ex-changeable basis $(e_n)_{n \in \mathbb{N}}$. Denote by $x(i)$ the ith coordinate of $x \in X$ [namely $x = \sum_{i=0}^{\infty} x(i)e_i$] and by x^* the unique element of X whose coordinates are $\{|x(i)|, i \in \mathbb{N}\}$, arranged in decreasing order. (x^* is the decreasing rearrangement of x.) Let $(x_n)_{n \in \mathbb{N}}$ be a bounded sequence in X.

(a) Show that there exists a subsequence $(y_n^*)_{n \in \mathbb{N}}$ of $(x_n^*)_{n \in \mathbb{N}}$ such that for all $i \in \mathbb{N}$, $(y_n^*(i))_{n \in \mathbb{N}}$ is convergent. Let $y^*(i)$ be its limit.

(b) Show that the series $\sum_{i=0}^{\infty} y^*(i)e_i$ is convergent. Let $y^* = \sum_{i=0}^{\infty} y^*(i)e_i$. Show that $(y^*(i))_{i \in \mathbb{N}}$ tends to 0 when $i \to +\infty$.

(c) Prove that for all $k \in \mathbb{N}$, there exist n_k and N_k in $\mathbb{N}$ such that if we define

$$z_k^* = \sum_{i=0}^{N_k} y_{n_k}^*(i)e_i \quad \text{and} \quad t_k^* = \sum_{i=N_k+1}^{\infty} y_{n_k}^*(i)e_i$$

we get

$$z_k^* \xrightarrow[k \to +\infty]{} y^* \quad \text{in } X$$
$$\| t_k^* \|_{\infty} = \text{Sup} \{| t_k^*(i) |, i \in \mathbb{N}\} \xrightarrow[k \to +\infty]{} 0$$

12. (See [Gu 4].) Let X be a separable stable space and σ a type on X, defined by a good sequence $(x_n)_{n \in \mathbb{N}}$ in X. We denote by $(e_n)_{n \in \mathbb{N}}$ the fundamental sequence of the spreading model defined by $(x_n)_{n \in \mathbb{N}}$. Let $K_1(\sigma)$ be the following set in $\mathcal{T}(X)$:

$$\{\tau \in \mathcal{T}(X)/\| \tau \| \le 1, \exists k \in \mathbb{N}, \exists \alpha_0, \ldots, \alpha_k \in \mathbb{R}^{k+1}$$

$$\text{such that } \tau = \alpha_0\sigma * \cdots * \alpha_k\sigma\}$$

(a) Show that if σ is an ℓ_p-type for some $p \in [1,+\infty[$, $K_1(\sigma)$ is relatively compact for uniform convergence on bounded sets of X.

(b) Suppose that $K_1(\sigma)$ is relatively compact for the topology of uniform convergence on bounded sets of X. Using Theorem

II.4.2, prove that, for all $\epsilon > 0$, there exists a subsequence $(x_{n_k})_{k \in \mathbb{N}}$ of $(x_n)_{n \in \mathbb{N}}$ such that:

$$\forall k \in \mathbb{N}, \; \forall (\alpha_0, \ldots, \alpha_k) \in \mathbb{R}^{k+1},$$

$$(1 - \epsilon) \left\| \sum_{i=0}^{k} a_i e_i \right\| \le \left\| \sum_{i=0}^{k} a_i x_{n_i} \right\| \le (1 + \epsilon) \left\| \sum_{i=0}^{k} a_i e_i \right\|$$

13. (See [Gu 4].) Let $L_{2k} = L_{2k}([0,1], \lambda)$, $k \in \mathbb{N}^*$.
    (a) Show that for all types σ on L_{2k}, there exist $\ell_i^\sigma \in L_{2k/i}$, $i = 1$, $\ldots$, $2k - 1$ such that:
       (i) $\forall x \in L_{2k}$, $\sigma(x)^{2k} = \| \sigma \|^{2k} + \| x \|^{2k} +$
       $\sum_{i=1}^{2k-1} C_{2k}^i \langle \ell_i^\sigma, x^{2k-i} \rangle$
       (ii) $\| \ell_i^\sigma \|_{2k/i}^{2k/i} \le \| \sigma \|_{L_{2k}}^{2k}$, $i = 1, \ldots, 2k - 1$
       where $\langle \, , \, \rangle$ denotes for every i the duality product between $L_{2k/i}$ and $L_{2k/2k-i}$. (Use Exercise 9(a) of Chapter III.) We recommend proving that ℓ_i^σ does not depend on the sequence $(x_n)_{n \in \mathbb{N}}$ which defines σ.
    (b) Show that $\sigma_n \xrightarrow[n \to +\infty]{} \sigma$ in $\mathcal{T}(L_{2k})$ if and only if

    $$\| \sigma_n \| \xrightarrow[n \to +\infty]{} \| \sigma \|$$
    $$\forall i = 1, \ldots, 2k - 1, \quad \ell_i^{\sigma_n} \xrightarrow[n \to +\infty]{} \ell_i^\sigma \quad \text{for } \sigma(L_{2k/i}, L_{2k/2k-i})$$

    (c) Show that if

    $$\| \sigma_n \| \xrightarrow[n \to +\infty]{} \| \sigma \|$$
    $$\forall i = 1, \ldots, 2k - 1, \quad \ell_i^\sigma \xrightarrow[n \to +\infty]{} \ell_i^\sigma \quad \text{in } L_{2k/i}$$

    then $\sigma_n \xrightarrow[n \to +\infty]{} \sigma$ for the topology of uniform convergence on bounded sets of L_{2k}. (See Exercise 12.)
    (d) If σ and τ are two types on L_{2k}, prove that for $i = 1, \ldots,$ $2k - 1$

    $$\ell_i^{\sigma * \tau} = \ell_i^\sigma + \ell_i^\tau + \sum_{j=1}^{i-1} C_i^j \ell_j^\sigma \ell_{i-j}^\tau$$
    $$\ell_i^{\alpha \sigma} = \alpha^i \ell_i^\sigma \quad (\alpha \in \mathbb{R})$$

    (e) Let $(x_n)_{n \in \mathbb{N}}$ be a good sequence in X, equivalent to the unit vector basis of ℓ_2, σ the type on L_{2k} defined by $(x_n)_{n\in\mathbb{N}}$, and $(e_n)_{n\in\mathbb{N}}$ the fundamental sequence of the spreading model on $(x_n)_{n\in\mathbb{N}}$. Using the following steps, show that $K_1(\sigma)$ (see Exercise 12) is relatively compact for the topology of uniform convergence on bounded sets of X:

   (i) Show that uniform convergence on bounded sets of X defines a metric topology on $\mathcal{T}(L_{2k})$.

  (ii) Show that if $(f_n)_{n\in\mathbb{N}}$ is a bounded sequence in the spreading model $\overline{\text{Span}}\,[e_i,\ i\in\mathbb{N}]$, such that

$$\forall n\in\mathbb{N},\qquad f_n = \sum_{i\in A_n}\alpha_i^n e_i,\ A_n = \{i_1,\ldots,i_{k_n}\}$$

$$(f_n)_{n\in\mathbb{N}}\ \text{converges in}\ \overline{\text{Span}}\,[e_i,\ i\in\mathbb{N}]$$

then the sequence of types $(\alpha_{i_1}^n\sigma * \cdots * \alpha_{i_k}^n\sigma)_{n\in\mathbb{N}}$ converges uniformly on bounded sets of X

 (iii) Let $(\tau_m)_{m\in\mathbb{N}}$ be a sequence in $K_1(\sigma)$. Show that there exist a subsequence $(\tau_{n_k})_{k\in\mathbb{N}}$ and two sequences $(\tau_k')_{k\in\mathbb{N}}$ and $(\tau_k'')_{k\in\mathbb{N}}$ such that

$$\tau_{n_k} = \tau_k' * \tau_k''$$
$(\tau_k')_{k\in\mathbb{N}}$ converges uniformly on bounded sets of X
$\tau_k'' = \alpha_1^k\sigma * \cdots * \alpha_{N_k}^k\sigma$ and $\text{Sup}\,\{|\,\alpha_1^k\,|\ \cdots\ |\,\alpha_{N_k}^k\,|\} \xrightarrow[k\to+\infty]{} 0$

(Use Exercise 11.) Deduce from this that $(\tau_{n_k})_{k\in\mathbb{N}}$ is relatively compact for uniform convergence on bounded sets of X if and only if $(\tau_k'')_{k\in\mathbb{N}}$ is like this. In the sequel, we can and will suppose that $(\tau_m)_{m\in\mathbb{N}}$ itself verifies

$$\forall m\in\mathbb{N},\qquad \tau_m = \alpha_1^m\sigma * \cdots * \alpha_{N_m}^m\sigma$$
$$\lim_{m\to+\infty}\text{Sup}\,\{|\,\alpha_1^m\,|,\ldots,|\,\alpha_{N_m}^m\,|\} = 0$$

 (iv) Show that we can also suppose, by taking a subsequence, that $(\sum_{i=1}^{N_m}|\,\alpha_i^m\,|^2)_{m\in\mathbb{N}}$ converges to the same α^2, $\alpha\neq 0$ when $m\to+\infty$. Deduce from this that $(\sum_{i=1}^{N_m}|\,\alpha_i^m\,|)_{m\in\mathbb{N}}$ tends to 0 when m tends to ∞ if $p > 2$.

(v) Under these hypotheses on $(\tau_m)_{m\in\mathbb{N}}$, show that for $i = 1,$
$\ldots, 2k - 1$

$$\begin{cases} \ell_i^{\tau_m} \xrightarrow[m\to+\infty]{} 0 & \text{if } i \text{ is odd, in } L_{2k/i} \\ \ell_i^{\tau_m} \xrightarrow[m\to+\infty]{} \dfrac{2j!}{2^j j!} (\alpha^2 \ell_2^\sigma)^j & \text{if } i = 2j, \text{ in } L_{2k/i} \end{cases}$$

(Remark that $\ell_1^\sigma = 0$.)

(f) Conclude question (e).

(g) Using Theorem IV.4.3, question e, and Exercise 12, state the property of bounded and weakly null sequences of L_{2k} that is proved here.

References

[A1]   Aldous, D., Unconditional bases and martingales in $L_p(F)$, *Math. Proc. Cambridge Philos. Soc. 85* (1979), 117–123.

[A2]   Aldous, D., Subspaces of L_1 via random measures, *Trans. Amer. Math. Soc. 267* (1981), 445–463.

[AA]   Alspach, D., and Argyros, S., Complexity of weakly null sequences. To appear in *Dissertationes Math.*

[Al]   Altschuler, Z., A Banach space with a symmetric basis which contains no ℓ_p or c_0, *Compositio Math. 35* (1977), 189–195.

[ACL]   Altschuler, Z., Casazza, P. G., and Bor-Luh Lin. On symmetric basic sequences in Lorentz sequence spaces, *Israel J. Math. 15* (1973), 140–155.

[An]   Andrew, A., Spreading basic sequences and subspaces of James' quasi reflexive space, *Math. Scand. 48* (1981), 108–118.

[ANZ]   Argyros, S., Negrepontis, S., and Zachariades, Th., Weakly stable Banach spaces, *Israel J. Math. 57* (1987), 68–88.

[Ar]   Arazy, J., On stability of unitary matrix spaces, *Proc. Amer. Math. Soc. 87* (1983), 317–321.

[Az] Azimi, P., and Hagler, J., Examples of hereditary ℓ_1 Banach spaces failing the Shur property, *Pacific J. Math. 122* (1986), 287–297.

[B] Baernstein, A., On reflexivity and summability, *Studia Math. 42* (1972), 91–94.

[Ban] Banach, S., Théorie des applications linéaires, Varsovie, (1932).

[Bas] Bastero, J., Embedding unconditional stable Banach spaces into symmetric stable Banach spaces, *Israel J. Math. 53* (1986), 373–380.

[BM] Bastero, J., and Mira, J. M., Stability of vector valued Banach sequence spaces, *Bull. Pol. Acad. Sci. Math. 34* (1986), 47–53.

[BR] Bastero, J., and Raynaud, Y., Quotient and interpolate spaces of stable Banach spaces, *Studia Math. 93* (1988).

[Be1] Beauzamy, B., Banach-Saks properties and spreading models, *Math. Scand. 44* (1979), 357–384.

[Be2] Beauzamy, B., Deux espaces de Banach et leurs modèles étalés, Dep. de Math. Univ. de Lyon I (1980).

[Be3] Beauzamy, B., *Introduction to Banach Spaces and Their Geometry*, "Notas de Mathematica," 68, North Holland, Amsterdam (1982).

[Be4] Beauzamy, B., Sous espaces de modèles étalés et structures cycliquement reproduites dans les espaces de Banach, *Bull. Sci. Math.* (1982).

[BeL] Beauzamy, B., and Lapreste, J. T., *Modèles étalés des espaces de Banach. Travaux en cours.* Hermann, Paris (1984).

[BeM] Beauzamy, B., and Maurey, B., Iteration of spreading models, *Ark. Math. 17* (1979), 193–198.

[BP1] Bessaga, C., and Pelczynski, A., On bases and unconditional convergence of series in Banach spaces, *Studia Math. 17*, (1958), 151–164.

[BP2]   Bessaga, C., and Pelczynski, A., A generalisation of results of R. C. James concerning absolute bases in Banach spaces, *Studia Math. 17* (1958), 165–174.

[Bo1]   Bourbaki, N., *Eléments de mathématiques. Topologie generale*, Hermann, Paris.

[Bo2]   Bourbaki, N., *Eléments de mathématiques. Espaces vectoriels topologiques*, Hermann, Paris.

[BCLT]   Bourgain, J., Casazza, P. G., Lindenstrauss, J., and Tzafriri, L., Banach spaces with a unique unconditional basis, up to permutation, *Mem. Amer. Math. Soc. 322* (1985).

[Br]   Brezis, H., *Analyse fonctionnelle, théorie et applications*, Masson, Paris (1983).

[Bru]   Brunel, A., Espaces associé à une suite bornée dans un espace de Banach, Séminaire Maurey-Schwartz, Ecole Polytechnique (1973–74).

[BS1]   Brunel, A., and Sucheston, L., On B-convex Banach spaces, *Math. Systems Theory 7* (1974), 294–299.

[BS2]   Brunel, A., and Sucheston, L., On J-convexity and ergodic super-properties of Banach spaces, *Trans. Amer. Math. Soc. 204* (1975), 79–90.

[Bu]   Bu, S., Deux remarques sur les espaces stables, *Compositio Math. 69* (1989), 341–355.

[Bur]   Burkholder, D. L., An elementary proof of an inequality of R. E. A. C. Paley, *Bull. London Math. Soc. 17* (1985), 474–478.

[C]   Canela, M. A., Stable quotients of ℓ_p, *Arch. Math. 44* (1985), 446–450.

[Ca]   Casazza, P. G., Tsirelson's spaces, *Proceedings, Research Workshop on Banach Spaces Theory*, Iowa City, 1981, University of Iowa Press (1982), 9–22.

[CJT]   Casazza, P. G., Johnson, W. B., and Tzafriri, L., Tsirelson's space, *Israel J. Math. 47* (1984), 81–98.

[CLL]   Casazza, P. G., Lin, B. L., and Lohman, R. H., On non-reflexive Banach spaces which contain no c_0 or ℓ_p, *Can. J. Math. 32* (1980), 1382–1389.

[CL]   Casazza, P. G., and Lohman, R. H., A general construction of spaces of the type of R. C. James, *Can. J. Math. 27* (1975), 1263–1270.

[CO]   Casazza, P. G., and Odell, E., Tsirelson's space and minimal subspaces, Longhorn Notes. University of Texas functional analysis seminar ((1982–83).

[CS]   Casazza, P. G., and Shura, T., *Tsirelson's Space*, Lecture Notes, 1363, Springer-Verlag, New York (1989).

[Ch]   Choquet, G., *Lectures on Analysis*, Vols. I, II, III, Math. Lecture Notes Series, Benjamin Cummings, Reading, Massachusetts.

[Da1]   Davie, A., The approximation problem for Banach spaces, *Bull. London Math. Soc. 5* (1973), 261–266.

[Da2]   Davie, A., The Banach approximation problem, *J. Approx. Theory* (1975), 392–394.

[Dav]   Davis, W. J., Embedding spaces with unconditional bases. *Israel J. of Math 20* (1975), 189–191.

[DFJP]   Davis, W. J., Figiel, T., Johnson, W. B., and Pelczynski, A., Factoring weakly compact operators, *J. Funct. Anal. 17* (1974), 311–327.

[D]   Day, M. M., *Normed Linear Spaces*, Springer-Verlag, New York (1973).

[DC]   Dacunha-Castelle, D., Sur un theorème de J. L. Krivine concernant la caracterisation des classes d'espaces isomorphes à des espaces d'Orlicz généralisés et des classes voisines, *Israel J. Math. 13* (1972), 261–276.

[DCK1]   Dacunha-Castelle, D., and Krivine, J. L., Application des ultraproduits à l'étude des espaces et des algèbres de Banach, *Studia Math. 41* (1972), 315–334.

[DCK2]   Dacuhna-Castelle, D., and Krivine, J. L., Sous espaces de L_1, *Israel J. Math. 26* (1977), 320–351.

[DSS]   Dean, D. W., Singer, J., and Sternbach, L., On shrinking basic sequences in Banach spaces, *Studia Math. 40* (1971), 23–33.

[De]   Deville, R., Geometrical implications of the existence of very smooth bump functions in Banach spaces, *Israel J. Math. 67* (1989), 1–22.

[Di1]   Diestel, J., *Geometry of Banach Spaces. Selected Topics*, Lecture Notes 485, Springer-Verlag, New York.

[Di2]   Diestel, J., *Sequences and Series in Banach Spaces*, Springer-Verlag, New York (1984).

[DU]   Diestel, J., and Uhl, J. J., *Vector Measures*. Math. Surv., Amer. Math. Soc. 15 (1977).

[Do]   Dor, L. E., On sequences spanning a complex ℓ_1-space, *Proc. Amer. Math. Soc. 47* (1975), 515–516.

[DS]   Dunford, N., and Schwartz, J. T., *Linear Operators*, Vols. I, II, III, Pure and Applied Math., Interscience, New York (1963).

[Dv]   Dvoretsky, A., Some results on convex bodies and Banach spaces, *Proceedings Symposium on Linear Spaces*, Jerusalem (1961), 123–160.

[El]   Elton, J., Extremely weakly unconditionally convergent series, *Israel J. Math. 40* (1981), 255–258.

[E1]   Enflo, P., A counter-example to the approximation problem in Banach spaces, *Acta Math. 130* (1973), 309–317.

[E2]   Enflo, P., On Banach spaces which can be given an equivalent uniformly convex norm, *Israel J. Math. 13* (1972), 281–288.

[F]   Farahat, J., Espaces de Banach contenant ℓ_1, d'après H. P. Rosenthal, Séminaire Maurey-Schwartz, Ecole Polytechnique, (1973–74).

[Fe]   Feller, W., *An Introduction to Probability Theory and Its Applications*, Vols. I and II, Wiley, New York. (1966).

[FJ]   Figiel, T., and Johnson, W. B., A uniformly convex space which contains no ℓ_p, *Compositio Math. 29* (1974), 179–190.

[GP]   Galvin, F., Prikry K. Borel sets and Ramsey's theorem, *J. of Symbolic Theory 38* (1973), 193–198.

[G]   Garling, D. J. H., *Stable Banach Spaces, Random Measures and Orlicz Function Spaces*, Lecture Notes in Math. 928 (1982).

[GM]   Ghoussoub, N., and Maurey, B., A non linear method for constructing certain basic sequences in Banach spaces, *Ill. J. Math. 34* (1990), 607–613.

[Go]   Godefroy, G., Metric characterisation of first Baire class linear forms and octahedral norms, *Studia Math. 95* (1989), 1–15.

[Gr]   Groh, U., Uniform ergodic theorems for identity preserving Schwarz maps on w*-algebras, *J. Operator Theory 11* (1984), 395–404.

[Gu1]   Guerre, S., Espaces quotient et stabilité, Note aux C.R.A.S., Paris t. 300, Serie I, 14 (1985), 485–487.

[Gu2]   Guerre, S., Sur les espaces stables universels, *Studia Math. 81* (1985), 221–229.

[Gu3]   Guerre, S., Application d'un résultat de J. Bourgain au problème des espaces stables universels, *Séminaire d'initiation à l'analyse*, Univ. Paris 6 (1984–85).

[Gu4]   Guerre, S., Types et suites symetriques dans L^p, $1 \le p < +\infty$ $p \ne 2$, *Israel J. Math. 53* (1986), 191–208.

[Gu5]   Guerre, S., Sur les suites presque échangeables dans L^q, $1 \le q < 2$, *Israel J. Math. 56* (1986), 361–380.

[GL1] Guerre, S., and Lapresté, J. T., Quelques propriétés des modèles étalés sur les espaces de Banach. *Ann. Inst. H. Poincaré Sect. B 16* (1980), 339–347.

[GL2] Guerre, S., and Lapreste, J. T., Quelques propriétés des espaces de Banach stables, *Israel J. Math. 39* (1981), 247–254.

[GLe] Guerre, S., and Levy, M., Espaces ℓ_p dans les sous-espaces de L_1, *Trans. Amer. Math. Soc. 279* (1983), 611–616; and Note aux C.R.A.S. Paris, 294 (1982), 167–170.

[GR] Guerre, S., and Raynaud, Y., On sequences with no almost symmetric subsequence, Longhorn notes, University of Texas Functional Analysis Seminar (1985–86).

[HLR] Haydon, R., Levy, M., and Raynaud, Y., *Randomly Normed Spaces*, Travaux en cours, Hermann (1991).

[HM] Haydon, R., and Maurey, B., On Banach spaces with strongly separable types, *J. London Math. Soc. 33* (1986), 484–498.

[HL] Haydon, R., and Lin, P. K., Ultrapowers of rearrangement invariant function spaces, Proceedings of the 1987 Iowa Workshop on Banach Spaces Theory, *Contemp. Math. 85* (1989), 253–280.

[H] Heinrich, S., Ultraproducts in Banach spaces theory, *J. Reine Angew. Math. 313* (1980), 72–104.

[Ha] Halmos, P. R., *Measure Theory*, Van Nostrand Reinhold (1950).

[Ho] Hoffmann-Jørgensen, J., Sums of independent Banach space valued random variables. *Studia Math.*, T.L11, (1979), 159–185.

[J1] James, R. C., A non reflexive Banach space isometric to its second conjugate, *Proc. Nat. Acad. Sc. U.S.A.* (1951), 174–177.

[J2] James, R. C., Reflexivity and the supremum of linear functionals, *Ann. Math. 66* (1957), 159–169.

[J3]   James, R. C., Bases and reflexivity of Banach spaces, *Ann. Math. 52* (1950), 518–527.

[J4]   James, R. C., Weak compactness and reflexivity, *Israel J. Math. 2* (1964), 101–119.

[J5]   James, R. C., Uniformly non square Banach spaces, *Ann. Math. 80* (1964), 542–550.

[J6]   James, R. C., Super-reflexive spaces with bases, *Pacific J. Math. 41* (1972), 409–419.

[J7]   James, R. C., Some self dual properties of normed linear spaces, Symposium on infinite dimensional topology, *Ann. Math. Studies 69* (1972), 159–175.

[J8]   James, R. C., Super-reflexive Banach spaces, *Can. J. Math. 24* (1972), 896–904.

[JR]   Johnson, W. B., and Rosenthal, H. P., On w*-basic sequences and their applications to the study of Banach spaces, *Studia Math. 43* (1972), 77–92.

[JRZ]   Johnson, W. B., Rosenthal, H. P., and Zippin, M., On bases, finite dimensional decompositions and weaker structures in Banach spaces, *Israel J. Math. 9* (1971), 488–506.

[JMST] Johnson, W. B., Maurey, B., Schechtman, G., and Tzafriri, L., *Symmetric Structures in Banach Spaces*, Mem. Amer. Math. Soc. *217* (1979).

[KP]   Kadec, M. I., and Pelczynski, A., Bases, lacunary sequences and complemented subspaces in the space L_p. *Studia Math. 21* (1962), 161–176.

[Ka]   Kakutani, S. Concrete representation of abstract L-spaces and the mean ergodic theorem. *Ann. Math. 42* (1941), 523–537.

[K1]   Krivine, J. L., Sous-espaces et cônes convexes dans les espaces L^p, Thèse d'Etat, Faculté des sciences de Paris (1967).

[K2]   Krivine, J. L., Sous-espaces de dimension finie des espaces de Banach réticulés, *Ann. Math. 104* (1976), 1–29.

[K3]   Krivine, J. L., Plongement de ℓ^p dans certains espaces de Banach, Séminaire d'Analyse Fonctionnelle, Ecole Polytechnique (1979–80).

[KM]   Krivine, J. L., and Maurey, B., Espaces de Banach stables, *Israel J. Math. 39* (1981), 273–295.

[Ko]   Köthe, G., *Topological Vector Spaces*, Springer-Verlag, New York (1969).

[Ku]   Kurakowski, K. *Topolgy*, Vol I, Academic Press, (1966).

[L]   Lacey, H. E., *The Isometric Theory of Classical Banach Spaces*, Springer-Verlag, New York, (1974).

[La1]   Lapreste, J. T., Suites écartables dans les espaces de Banach, Séminaire d'Analyse Fonctionnelle, Ecole Polytechnique (1977–78).

[La2]   Lapreste, J. T., Suites asymptotiquement inconditionnelles, Séminaire d'Analyse Fonctionnelle, Ecole Polytechnique (1978–79).

[Le]   Lemberg, H., Nouvelle démonstration d'un théorème de J. L. Krivine sur la finie représentation de ℓ^p dans un espace de Banach, *Israel J. Math. 39* (1981), 341–348.

[Lev]   Levy, M., L'espace d'interpolation réel $(A_0, A_1)_{\vartheta, p}$ contient ℓ_p, Note aux C.R.A.S., Paris, 289 (1979), 675–677.

[LR]   Levy, M., and Raynaud, Y., Ultrapuissances de $L^p(L^q)$, Note aux C.R.A.S., Paris, 299 (1984), 81–84.

[LiR]   Lindenstrauss, J., and Rosenthal, H. P., The $\mathscr{L}^p$-spaces, *Israel J. Math. 7* (1969), 325–349.

[LT]   Lindenstrauss, J., and Tzafriri, L., *Classical Banach Spaces I and II*, Lecture Notes, 338, Springer-Verlag, New York (1977).

[M1]   Maurey, B., Le système de Haar, Séminaire Maurey-Schwartz, Ecole Polytechnique (1974–75).

[M2]   Maurey, B., Toute espace L^1 contient un ℓ^p-d'après D. Aldous, Séminaire d'analyse fonctionnelle, Ecole Polytechnique (1979–80).

[M3]   Maurey, B., Types and ℓ^1-subspaces, Longhorn Notes, University of Texas Functional Analysis Seminar (1982–83).

[M4]   Maurey, B., Sous-espaces ℓ^p des espaces de Banach, Séminaire Bourbaki, 608 (1982–83).

[MP]   Maurey, B., and Pisier, G., Série de variables aléatoires indépendantes et propriétés géométriques des espaces de Banach, *Studia Math.* T LVIII (1976), 45–90.

[MR]   Maurey, B., and Rosenthal, H. P., Normalized weakly null sequence with no unconditional subsequence, *Studia Math.* T LXI (1977), 77–98.

[MS]   Maurey, B., and Schechtmann, G., Some remarks on symmetric basic sequences in L^1, *Compositio Math 38* (1979), 67–76.

[Mi]   Milman, V. D., Geometric theory of Banach spaces, English translation, *Russian Math. Surv. 25* (1970), 111–170.

[MiS]   Milman, V. D., and Schechtmann, G., *Asymptotic Theory of Finite Dimensional Normed Spaces*, Lecture Notes in Math., 1200, Springer-Verlag, New York. (1986)

[MiSh] Milman, V. D., and Sharir, M., A new proof of the Maurey-Pisier theorem, *Israel J. Math. 33* (1979), 73–87.

[N-W]   Nash-Williams, C. St. J. A. On well ordering transfinite sequences. *Proc. Cambridge Phil. Soc.*, 61, (1965), 33–39.

[N]   Neveu, J., *Bases mathématiques du calcul des probabilités*, Masson, Paris (1970).

[O1]   Odell, E., Applications of Ramsey theorem to Banach spaces theory, University of Texas at Austin Press (1981), 379–404.

[O2]   Odell, E., A normalized weakly null sequence with no shrinking subsequence in a Banach space not containing ℓ_1, *Compositio Math. 41* (1980), 287–295.

[OR]   Odell, E., and Rosenthal, H. P., A double dual characterization of Banach spaces containing ℓ_1, *Israel J. Math. 20* (1975), 375–384.

[OW]   Odell, E., and Wage, W., Weakly null normalised sequences equivalent to unit basis of c_0, University of Texas at Austin.

[Pa]   Paley, R. E. A. C., A remarkable series of orthogonal functions, *Proc. London Math. Soc. 34* (1932), 241–264.

[Pe1]   Pelczynski, A., On the impossibility of embedding of the space L in certain Banach spaces, *Colloq. Math. 8* (1961), 199–203.

[Pe2]   Pelczynski, A., Projections in certain Banach spaces, *Studia Math. 19* (1960), 209–228.

[Pe3]   Pelczynski, A., A connection between weakly unconditional convergence and weak completeness of Banach spaces, *Bull. Acad. Pol. Sci. 6* (1958), 251–253.

[Pe4]   Pelczynski, A., Universal bases. *Studia Math. 32* (1969), 247–268.

[PS]   Pelczynski, A., and Szlenk, W., An example of a non shrinking basis, *Rev. Roum. Math. Pures Appl. 10* (1965), 961–966.

[P1]   Pisier, G., Bases, lacunary sequences and complemented subspaces in space L^p, d'après Kadec-Pelczynski, Séminaire Maurey-Schwartz, Ecole Polytechnique (1972–73).

[P2]   Pisier, G., Sur les espaces qui ne contiennent pas de ℓ_n^∞ uniformément, Séminaire Maurey-Schwartz, Ecole Polytechnique (1972–73).

[P3]   Pisier, G., Sur les espaces qui ne contiennent pas de ℓ_n^1 uniformément, Séminaire Maurey-Schwartz, Ecole Polytechnique (1972–73).

[P4]   Pisier, G., Type des espaces normés, Séminaire Maurey-Schwartz, Ecole Polytechnique (1972–73).

[P5]   Pisier, G., Martingales with values in uniformly convex spaces, *Israel J. Math. 20* (1975), 326–350.

[P6]   Pisier, G., Les inégalités de Khintchine-Kahane d'aprés C. Borell, Séminaire Maurey-Schwartz, Ecole Polytechnique (1977–78).

[P7]   Pisier, G., Une propriété de stabilité des espaces ne contenant pas ℓ^1, Note aux C.R.A.S., Paris, 286 (1978), 747–749.

[P8]   Pisier, G., Holomorphic semi-groups and the geometry of Banach spaces, *Ann. Math. 115* (1982), 375–392.

[P9]   Pisier, G., *Probabilistic Methods in the Geometry of Banach Spaces*, Springer Lecture Notes in Math. 1206 (1986), 167–241.

[R]   Ramsey, F. P., On a problem of formal logic, *Proc. London Math. Soc. 2-30* (1929), 264–286.

[Ra1]   Raynaud, Y., Espaces de Banach superstables, distances stables et homéomorphisme uniformes, *Israel J. Math. 44* (1983), 33–52.

[Ra2]   Raynaud, Y., Deux nouveaux exemples de Banach stables, Note aux C.R.A.S., Paris, t.292, Série I (1981), 715–717.

[Ra3]   Raynaud, Y., Stabilité des espaces C_E, Séminaire de géométrie des espaces de Banach, Univ. Paris 7 (1982–83).

[Ra4]   Raynaud, Y., Sur les sous-espaces de $L^p(L^q)$, Séminaire de géométrie des espaces de Banach, Univ. Paris 6-Paris 7 (1984–85).

[Ro]   Ropars, Y., Modèles étalés des espaces de Banach, Thèse de 3ème cycle, Univ. Paris 7 (1983).

[Ros1]   Rosenthal, H. P., On subspaces of L^p ($p > 2$) spanned by sequences of independant random variables, *Israel J. Math. 8* (1970), 273–303.

[Ros2]   Rosenthal, H. P., On subspaces of L^p, *Ann. Math. 97* (1973), 344–373.

[Ros3]   Rosenthal, H. P., A characterisation of Banach spaces containing ℓ_1, *Proc. Nat. Acad. Sci. U.S.A. 71* (1974), 2411–2413.

[Ro4]   Rosenthal, H. P., Some remarks concerning unconditional basic sequences, Longhorn notes, Texas Functional Analysis Seminar (1982–83).

[Ros5]   Rosenthal, H. P., Double dual types and the Maurey characterization of Banach spaces containing ℓ_1, Longhorn notes, Texas Functional Analysis Seminar (1983–84).

[Ros6]   Rosenthal, H. P., The unconditional basic sequence problem, *Contemp. Math. 52* (1986), 70–98.

[Ros7]   Rosenthal, H. P., On a theorem of J. L. Krivine concerning block finite representability of ℓ_p in general Banach spaces, *J. Funct. Anal. 28* (1978), 197–225.

[Ros8]   Rosenthal, H. P., Some aspects of the subspace structure of infinite dimensional Banach spaces, to appear in *Approximation Theory and Functional Analysis*, Academic Press.

[Ros9]   Rosenthal, H. P., Weakly independent sequences and the Banach-Saks property. Proc. of the Durham Symposium (1975).

[Ru]   Rudin, W., *Real and Complex Analysis*, McGraw-Hill, New York (1966).

[S]   Schachermayer, W., The class of Banach spaces, which do not have c_0 as a spreading model, is not hereditary, *Ann. Inst. H. Poincaré 19* (1983), 1–8.

[Sc]   Schutt, C., Lorentz spaces that are isomorphic to subspaces of L_1, *Trans. Am. Math. Soc. 314* (1989), 583–595.

[Si]   Singer, I., *Bases in Banach spaces I and II*, Springer-Verlag, New York (1970).

[So]   Sobczyk, A., Projection of the space m on its subspace c_0, *Bull. Am. Math. Soc. 47* (1941), 938–947.

[St1]   Stern, J., Propriétés locales et ultrapuissances d'espaces de Banach, Séminaire Maurey-Schwartz, Ecole Polytechnique (1974–75).

[St2]   Stern, J., Some applications of model theory in Banach space theory, *Ann. Math. Logic 9* (1976), 49–122.

[St3]   Stern, J., Ultrapowers and local properties of Banach spaces, *Trans. Amer. Math. Soc. 240* (1978), 231–252.

[Sto]   Stone, M. H., Applications of Boolean rings to general topology, *Trans. Amer. Math. Soc. 41* (1937), 375–481.

[Sza]   Szarek, S., A Banach space without a basis which has the bounded approximation property, *Acta Math. 159* (1987), 81–98.

[Sz]   Szlenk, W., Sur les suites faiblement convergentes dans l'espace L, *Studia Math. 25* (1965), 337–341.

[T]   Tsirelson, B. S., Not every Banach space contains ℓ_p or c_0, *Funct. Anal. Appl. 8* (1974), 138–141.

[V]   Veech, W. A., Short proof of Sobczyk's theorem, *Proc. Amer. Math. Soc. 28* (1971), 627–628.

[Y]   Yoshida, K., *Functional Analysis*, Springer-Verlag, New York, (1971).

[Z1]   Zippin, M., On perfectly homogeneous bases in Banach spaces, *Israel J. Math. 4* (1966), 265–272.

[Z2]   Zippin, M., A remark on bases and reflexivity in Banach spaces, *Israel J. Math. 6* (1968), 74–79.

[Z3]   Zippin, M., The separable extension problem, *Israel J. Math. 26* (1977), 372–387.

Index

Basis, 1–2, 7, 14–15, 17–18,
  21–22, 28, 36, 39,
  51–53, 66–67, 98, 104
basis constant, 6, 8, 10,
  12–14, 16–19, 21, 42,
  52, 84
block basis, 14–15, 17, 26,
  28, 37–38, 40–41,
  53–54, 56–57
boundedly complete basis,
  18–21, 36–38
coordinates on a basis, 1–2,
  4, 15, 76, 98, 185
monotone basis, 6–7, 32, 67
normalized basis, 6–7, 37,
  56–57, 64–65, 68–69,
  95, 124
shrinking basis, 18–21,
  36–37, 67–69

summing basis, 6, 26, 36
support on a basis, 1–2, 16,
  39–40, 56, 64–65, 70,
  99–100
unconditional basis, 22,
  26–28, 30, 32, 35–36,
  38–39, 56, 66–67, 106,
  163, 185
  K-sign-unconditional
  basis, 93, 95–99,
  121–122, 124, 127
  K-unconditional basis, 26,
  37–38, 64, 84, 87, 89,
  93, 95–97, 99–100, 102,
  121
unconditional basis
  constant, 26, 30, 36, 58
unit vector basis (unit
  vectors), 6, 15–17,

Basis (*continued*)
　　20–21, 26, 30, 37–39,
　　41–43, 46, 48, 51, 54,
　　64–65, 67, 69, 76, 81,
　　86, 89–90, 96, 99–102,
　　104–107, 109, 127, 129,
　　146, 162–166, 173, 183,
　　187
Bessaga–Pelczynski theorem,
　　41
Biorthogonal functionals
　　(forms, system), 17, 52,
　　55, 66–68
Brunel–Sucheston theorem,
　　71, 78

Dacunha–Castelle–Krivine
　　theorem, 71, 73

Enflo's theorem, 7, 53
$\mathscr{E}$-net, 7–9, 76, 90, 114

Fourier transform, 167–168
Functions
　first Baire class functions,
　　113–114, 125
　Haar functions (system), 6,
　　32–34, 67
　Rademacher functions, 27,
　　30, 73, 146–148, 161,
　　164, 183
　separately continuous
　　functions, 112–113,
　　117–118, 128
　support of a function,
　　145–146, 166

Helly's condition, 59

James' sequence, 58, 130
James' space, 66–67, 105
James' theorem, 21, 36, 39, 96

Kadec–Pelczynski theorem,
　　162, 171
Khintchine's inequalities, 30,
　　147, 152, 161
Khintchine–Kahane
　　inequalities, 147–148,
　　152
Krivine's theorem, 92,
　　103–104, 172
Krivine–Maurey theorem, 90,
　　108, 122

Lattice, 66, 153–154
　sublattice, 155

Magic formula, 174–179, 182
Maurey–Pisier theorem, 172

Nontrivial ultrafilter, 73, 78,
　　82, 95, 99, 108–109, 184

Pelczynski's decomposition
　　method, 55
Pelczynski's theorem, 26–27

Rademacher cotype, 73, 146,
　　151–152, 167, 172
Rademacher type, 72–73, 146,
　　151–152, 164, 167,
　　172–173, 182
Ramsey's theorem, 43, 46, 78

Random variables, (r.v.)
148–149, 167, 169
independent and identically
distributed r.v. (i.i.d.),
27, 139, 147, 161, 168,
169, 183
(standard) gaussian r.v., 161
(standard) q-stable r.v., 167,
169
Rosenthal's theorem, 43, 84,
89, 183

Schauder basis (*see* Basis)
Schur property, 69
Sequence
basic sequence, 1, 7–14,
17–18, 22, 42, 84, 87,
89, 121, 127
block-finitely representable
sequence, 81–82, 92, 95,
99
blocks on a sequence, 14,
16–17, 29, 64–65,
68–69, 82, 95–96, 99,
103–104, 124, 126,
128–129
equivalent sequence, 11,
14–17, 30, 37–38, 42,
46–48, 56–57, 65,
67–69, 84, 86, 89, 95,
104–107, 126–128
C-equivalent sequence,
13, 55, 81, 84, 90, 93,
95–96, 99–103, 121,
126, 128, 162, 165–166
exchangeable sequence

Sequence (*continued*)
K-exchangeable sequence,
121, 185
fundamental sequence,
79–80, 82, 86–87,
89–90, 95, 104–107,
121–122, 124, 127, 185,
187
good sequence, 82–83,
86–87, 90, 92, 95, 104,
106–107, 121, 126–127,
185
extraction of a good
sequence, 82
invariant by spreading
sequence (I.S.
sequence), 81, 83–85,
95–99, 106
I.S. sequence over the
space, 81
peak sequence, 146, 162,
183
symmetric sequence
almost symmetric
sequence, 183
K-symmetric sequence,
121
Set
admissible family of sets,
63–65
independent sets, 46, 49, 51
norming set, 10–11
p-equi-integrable set,
139–146, 171, 183–184
Space (Banach space)
complexification of a space,
93, 98

Space (*continued*)

$C(\Omega)$-space, 46, 48, 67, 113, 115

c_0-space, 6, 15, 21, 26, 30, 32, 35–36, 38–43, 51, 53–57, 63, 65–66, 68–71, 81, 90, 92, 96, 99–101, 103–105, 107, 109, 123, 129–130, 133, 161, 185

finite representability of space, 72–74, 77, 80, 103–104, 170, 172–173, 181

ℓ_1-space, ℓ_1^n-space, 15–17, 20–21, 35–40, 43, 46–48, 51, 64–69, 72, 76, 89, 104–106, 127, 129, 130, 171, 183, 185

ℓ_p-space, ℓ_p^n-space, $\ell_p(X)$-space, 6, 15, 26, 35, 41, 51–56, 63, 65, 69–71, 81, 90, 92, 99, 100, 102–105, 108, 113, 119–120, 122–123, 126, 128–131, 133–134, 138, 146, 148, 151–153, 161–167, 169–173, 181–184, 186

ℓ_∞-space, ℓ_∞^n-space, 7, 16, 36, 39, 68–70, 73–74, 76, 183

L_1-space, $L_1[0, 1]$, $L_1(\Omega, \Sigma, \mu)$, $L_1(X)$, 20, 26–28, 30, 32, 36, 108, 122, 130, 140–145, 171, 173–174, 177, 182–183

Space (*continued*)

L_p-space, $L_p[0, 1]$, $L_p(\Omega, \Sigma, \mu)$, $L_p(X)$, 6, 26, 32, 34, 108, 113, 119–120, 131–136, 138–140, 145–148, 151, 154, 157–158, 160–167, 169–173, 175, 181, 183–184, 186–188

L_∞-space, $L_\infty[0, 1]$, 7, 28, 143–144, 183

reflexive space, 20–21, 36, 39, 51, 58–59, 61, 63, 68, 72–73, 104, 113, 115–116, 120, 127, 130, 132, 140, 161, 171, 183, 185

super reflexive space, 104, 160, 185

stable space, 108–109, 111–113, 115–123, 126–127, 129–131, 133, 171, 182, 184

super stable space, 129, 160, 184

weakly stable space, 129

uniformly convex space, 72–73, 104

weakly sequentially complete space, 36, 38–39, 51, 127, 129, 144

Spreading model, 71, 78–80, 82–83, 86–87, 89–90, 95, 104–105, 120–124, 127, 130, 185, 187

spreading model isometrically ℓ_p or c_0 over X, 90

Space (*continued*)
spreading model over X, 79,
87, 89–90, 105
Subsequence splitting lemma,
145
Super property, 73, 104
$\sigma(X, X^*)$-Cauchy (weak
Cauchy), 9–10, 36, 43,
48, 84, 86–87, 89, 105,
127, 129, 144
$\sigma(X, X^*)$-convergence (weak
convergence), 9, 10, 15,
19, 27, 36, 39, 41, 43,
59, 63, 67, 69, 84–89,
105–106, 115, 127, 129,
140–141, 144, 171, 177,
183, 185–186
$\sigma(X^*, X)$-convergence (weak*
convergence), 10, 48,
68–69, 184
$\sigma(X, X^*)$-limit(weak limit), 19,
43, 67, 84, 105, 120,
175–177, 179, 181

Transpose, 18, 20
Tchebichev's inequality, 140,
150, 172
Tsirelson's space, 63–64, 66,
105

Type on a space, 109–11,
120–122, 124, 126,
129–130, 175–178, 181,
183–187
conic class of types,
123–126, 128, 173,
181–182
convolution product of
types, 11, 128, 180
c_0-type, 123, 125–126
extension of types, 177–181
ℓ_p-type, 123–126, 129–130,
173, 180–185
positive type, 179–182
p-approximating sequence of
types, 124–125
realized type, 110, 112, 122
restriction of types, 173,
177–181
symmetric type, 122–123,
127–128, 179–180, 184

Ultrapower, 71–74, 77–78, 94,
133, 153–154, 160, 170,
172–173, 178, 183–184
reiteration of ultrapowers,
78
Ultraproduct, 74
$\mathcal{U}$-tree of trunk s, 44–46